A Defense of Computational Physics

by

Patrick J. Roache

hermosa
publishers

New Mexico, USA
http://www.hermosa-pub.com/hermosa

Printed-On-Demand by
CreateSpace, an Amazon.com Company

Patrick J. Roache

Copyright 2012 Patrick J. Roache
ISBN-13: 978-1469975085
ISBN-10: 1469975084

Library of Congress Control Number 2011962694

BISAC: Science / Philosophy and Social Aspects

Printed in the United States of America
Revised Printing July 2013

All rights reserved. No part of this book may be reproduced or transmitted in any form or by any means, electronic or mechanical, including photocopying, recording or by any information storage and retrieval system, without written permission from the copyright holder, except for brief quotations in a review.

Warnings - Disclaimer

This book is intended to provide information in regard to the subject matter covered. Reprinted material is quoted with permission where required, and sources are indicated. It is sold with the understanding that the publisher and author are not engaged in rendering legal, accounting, professional engineering or other professional services. If legal or other expert assistance is required, the services of a competent professional should be sought. Efforts have been made to make this book as complete and accurate as possible. However, there may be mistakes both typographical and in content. Therefore, the text should be used only as a general guide, not as the ultimate source of information. The purpose of this book is to educate. The publisher and author have neither liability nor responsibility to any person or entity with respect to any loss or damage caused, or alleged to be caused, directly or indirectly by the information contained in this book.

Errata and Addenda

Errata and Addenda for this book will be posted on
www.hermosa-pub.com/hermosa.

A Defense of Computational Physics

Cover Illustration

The cover illustration is the Ptolemaic geocentric model of the solar system as depicted by the Portuguese mapmaker and cosmographer Bartolomeu Velho in his *Cosmographia* (1568).

Patrick J. Roache

Other books by the author

Computational Fluid Dynamics, 1972, 1976
also translated into Japanese, Russian, and Chinese

Elliptic Marching Methods and Domain Decomposition, 1995

Verification and Validation in Computational Science and Engineering, 1998

Fundamentals of Computational Fluid Dynamics, 1998

Fundamentals of Verification and Validation, 2009

with Catharine Stewart-Roache
At Sea At Sixty, 1999

A Defense of Computational Physics

Dedication

To all conscientious modelers

Table of Contents

Preamble 8

Chapter 1
Introduction: The Need for a Defense 9
    References for Chapter 1 16

Chapter 2
Popper's Non-Verifiability vs.
Computational Validation 17

2.1	Background and Motivation	18
2.2	*Popular Popper Précis*	19
2.3	Objectives	21
2.4	Critique Level (A). Problems for the Philosophy of Science	22
2.4.1	Popper Invented *Falsificationism*, not Falsifiability	22
2.4.2	Demarcation	23
2.4.3	Changing Terminology	26
2.4.4	Non-Computational Difficulties with Popper's *Falsificationism*	28
2.5	Critique Level (B). Empirical Data on Science Practice: *Falsificationism* Falsified	31
2.6	Critique Level (C). Popper's Philosophy applied to Validation in Computational Physics Modeling	34
2.6.1	Critique Level (C-1). Common Sense	35
2.6.2	Critique Level (C-2). Are Models Equivalent to Theories?	36
2.6.3	Critique Level (C-3). Inclusion of Popper's *Numerical Universality* in Validation	37

2.6.4	Critique Level (C-4). Popper's *Truth* vs. Validation *Accuracy*	38
2.7	A Provocative Example of Validation of a Computational Model	45
2.8	Not All Models are "Wrong"!	49
2.9	Limited Domain of Validation	50
2.10	Popper and "Normal Science" of Kuhn	51
2.11	Contrasting Characteristics of Popper's *Falsificationism* vs. Computational Model Validation	53
2.12	Summary	55
2.13	Last Gasp	56
	Acknowledgements for Chapter 2	58
	References for Chapter 2	59

Chapter 3
Validation: What Does It Mean ? 63

3.1	Introduction	64
3.2	History of the Definition	65
3.3	Issue #1. Acceptability Criteria (Pass/Fail)	67
3.4	Issue #2. Necessity for Experimental Data	71
3.5	Issue #3. Intended Use	72
3.6	Recommended Interpretation and Alternative Description	74
3.7	Calibration is Not Validation	77
3.8	Implications for Contractual and Regulatory Requirements	77
3.9	Addendum: Expanded Definition of Validation	78
	Acknowledgements for Chapter 3	80
	References for Chapter 3	81

Glossary	83
Author Vita	85
End Notes	89

Patrick J. Roache

Preamble

This book is about philosophy of science. The non-technical reader will be relieved to know that no mathematical skills are required to read this book; the subject is much more about words than equations.

Chapter 1

Introduction: The Need for a Defense

Patrick J. Roache

While discussing the politicized topic of climate modeling, a physician friend who knew my professional interests asked me, "Are science models accurate?" My brief response was "Yes, roughly as accurate as the science behind the models." Whatever climate science is available can only be *used* in climate models.

To a great extent, *modeling* is how applied science is done in the 21st century. I hold this opinion even though my resume shows that I actively opposed the trend beginning in the mid-1970's towards excessive reliance on and naive credibility in models, sometimes to the detriment (scrapping) of good experimental facilities.

It is widely acknowledged that the emergence, over the last half century or so, of the broad field of computational science, mathematics, and engineering has revolutionized scientific and engineering practice. The shorter term *computational physics* is meant herein to be interpreted in the broadest possible sense, covering computational fluid dynamics (or CFD, my own specialty), computational chemistry, computational structural mechanics, electromagnetics, ecosystem modeling, climate modeling, etc. Examples of the widespread influence abound. Every major automobile manufacturer uses CFD. As of 2006, Hutton [1] cited estimates for the worldwide number of commercial CFD code users of 25-30,000, including 2/3 of the Fortune 500 companies. *Every* new automobile, airplane, and ship is designed with computational models of structural mechanics.

Why then is there a need to "defend" computational physics? In my view, the threats come from two sides, both concerned with standards for quality work. One sets standards that are too low (often just plain sloppy), the other too demanding (impossibly so, literally). On the low side,

A Defense of Computational Physics

there is now available wide access to commercial codes; sometimes these are of questionable accuracy, but even the highest quality codes can produce bad answers when poor work is done by unqualified personnel. This threat from the low quality side is the more serious and important one.

On the other side, the less obvious threat comes from the realm of philosophy of science, asserting that Validation of models is not possible: that Validation of computational physics models is inherently, essentially, unqualifiedly impossible. It might seem that this posture of setting high standards for computational work would encourage quality work, but in fact it poses a threat that is potentially as pernicious as that of sloppy standards. By stating that Validation of models is inherently impossible, this posture opens the door to low quality work. Since no one can do it right, then we are all in the same category, and anyone has an excuse for doing sloppy work. This latter posture is all the more pernicious because it is a *subtle* threat; seeming to set high standards, it actually makes excuses for low standards. This situation recalls the morbid condition of "scrupulosity" in classical moral theology, in which a person "sees and fears sin where there is none" and is "unable to formulate a practical judgment concerning the morality of action" [2]. Scrupulosity has long been recognized as dangerous and more difficult to overcome than more obvious character faults.

I have worked and written extensively in the field of Verification and Validation, with the objective of encouraging and enabling high quality computational work. But this booklet is different. It primarily addresses the scrupulosity issue. Chapter 2 addresses the threat that comes from the *falsificationism* philosophy of the late Karl Popper. This threat *should* easily be dismissed with the application of a little common sense, and so it *should not* be much of a problem or threat for practitioners. But we have empirical evidence that on occasion there have been troublesome obstacles placed in the practitioners' path by dogmatic citation of Popper's philosophy of *falsificationism*. Such situations provide the motivation for the defense presented in Chapter 2.

Essential to this presentation in Chapter 2 are the definitions of and distinctions between Verification and Validation. This semantic issue is further addressed in Chapter 3, devoted to the definition(s) of Validation.

I hope that the reader will not scoff at such "mere semantics." If you have a project contract that specifies you must use a "Verified and Validated code," it helps to know what this means, to all parties. If a regulator (e.g., the U.S. Environmental Protection Agency, or the Nuclear Regulatory Commission) can disallow your computational modeling and shut down your multi-year project, all based on the definitions propounded by a philosopher who wrote his main theses before the advent of modern computers and who never ran a computer code in his life, you may find yourself seriously concerned with "mere semantics." And if you find yourself in public hearings defending your position from intellectually dishonest attacks of stakeholders, you might appreciate the ammunition provided herein.

An essential ingredient of semantic arguments is the recognition of the paucity of language, even for eclectic English with the largest vocabulary of any language. The same words have different meanings in different contexts because there are simply not enough words to go around to cover all our needs. This is the reason I prefer to capitalize the words Verification and Validation. (This practice is not standard.) I do so to emphasize that, herein, these are technical terms defined in a technical context. Not even the basic meanings, let alone subtle distinctions, will be discerned by consulting a common-use dictionary.

> **"All our philosophy is a correction of the common usage of words," says Lichtenberg. Many of the quarrels and mistakes occurring in the course of scientific advance could be avoided if this remark were always remembered.**[A]

Until one has tried to hammer out in committee a widely acceptable definition of a technical term, even in the ideal situation of working with intelligent and honest people (as I have been fortunate to experience), I doubt that one can appreciate how difficult these semantic issues can be. And in fact, although I recognize the need for definitions, I inherently distrust them and prefer to use an expanded and multi-level description of words and concepts, rather than bounded legalistic definitions.

Since my intention is to counter this threat that comes from the "Validation is impossible" side, readers unfamiliar with my previous work may be inclined to dismiss this present work and myself as a simplistic "denier" of any problems with credibility of computations. I wish to emphasize that I have spent a large part of my career intensely involved with issues of quality and standards for computational solutions. For decades I personally have been strongly identified with the more skeptical end of the spectrum, and may even claim to have been at the historical forefront of significant discussions of the limitations of computational physics, as opposed to over-selling its capabilities. I have been confronted for having unrealistically high standards of accuracy and code Verification, have regularly complained in papers and committees and books and short courses about unsubstantiated claims of software quality, and have received professional awards for my work on *conservatively* quantifying numerical uncertainty. To document my values, without claiming any quality level for my own work, I present a timeline of my involvement on the high quality side in an endnote.[B] So I am not one to take Verification and Validation lightly, especially for high consequence projects. Perhaps this timeline will help make credible my claim that Popper's *falsificationism* is not, or rather should not be, a genuine problem for computational physics.

Chapter 2 will present a multi-level attack on *falsificationism,* first generally, then specifically as applied to computational physics. If a reader accepts the arguments at the first level, he may not bother to read the other levels. But these still may be worth reading if he is vitally interested in the subject, e.g. a supervisor or regulator may not accept the first level argument, in which case he would consider retrenching to level 2, etc. The whole subject is not inherently difficult (certainly not like some dense mathematical issue).

The tone of Chapter 2 is unpleasantly negative, because that is what the subject requires, unfortunately. Chapter 3 provides a somewhat more constructive perspective on Validation.

One further explanation is warranted before proceeding to the core of this booklet. As the reader will recognize, the subject is not typically covered in research journals in computational physics. I made one attempt to publish a shorter version of Chapter 2 in such a journal, but the paper was understandably rejected as inappropriate (along with some other justifiable criticisms of substance). The reviewers suggested publication in a philosophy journal.

A Defense of Computational Physics

I considered submission to the *Journal of the Philosophy of Science Association*, but I found the style of communication and submission to be awkward compared to journals with which I am familiar. Also, I do not have the credentials to establish authority in philosophical circles, so ultimate rejection would not be surprising, and delay of publication by a year or more would be a certainty. Most importantly, even if I succeeded in publishing in the philosophy journal, I would not be reaching the audience intended, which is the computational scientists "in the trenches." Hence the publication of this modest booklet.

I had the pleasure of presenting the material of Chapter 2 in a seminar at the University of Notre Dame, my beloved alma mater, while serving as Visiting Professor of Aerospace and Mechanical Engineering for the fall semester 2010, and again as a Keynote Address at the ASME Symposium on Verification and Validation, held 2-4 May 2012 in Las Vegas, Nevada. The PowerPoint slide presentation is available to the public at no charge at *www.hermosa-pub/hermosa*.

References for Chapter 1

1. Hutton, A. G. (2006), "Quality and Reliability in CFD," Plenary Presentation, FEDSM2006, ASME Joint U.S. European Fluids Engineering Summer Meeting, 17-20 July 2006, Miami, Florida, U.S.A.
2. McBrien, R. P. (1995), General Ed., "The HarperCollins Encyclopedia of Catholicism," p. 1179.

Chapter 2

Popper's Non-Verifiability vs. Computational Validation

Sir Karl Popper is often considered the most influential philosopher of science of the first half (at least) of the 20th century. His assertion that true science theories are characterized by falsifiability only has been used to discriminate between science and pseudo-science, and his assertion that science theories cannot be verified but only falsified has been used to categorically and pre-emptively reject claims of realistic Validation of computational physics models. Both of these assertions will be challenged, as well as the applicability of the second assertion to modern computational physics models such as climate models, even if it were considered to be correct for scientific theories.

Patrick J. Roache

2.1 Background and Motivation

Arguably the two most influential philosophers of science of the 20th century are Karl Popper (1902-1994) and Thomas Kuhn (1922-1996). They provide *markedly* different approaches. Kuhn is insightful, realistic, and pragmatic. His iconic contributions to science and to the broader culture include the terms and concepts of "paradigm" and "paradigm shift" and "theory-laden experiments," which he developed during the unusual process of studying *what scientists actually do*, i.e. an historical approach to philosophy of science. Unlike Popper, Kuhn was not just a philosopher of science but a scientist, a Harvard PhD physicist. His approach and philosophy are not problematical to computational physics; not so for that of Karl Popper.

Popper's most well-known work is *The Logic of Scientific Discovery* [1], the summary English edition published in 1963. His other well known book is [2]. He is very popular with scientists themselves, more so than other 20th century philosophers such as Peirce, Quine, and Feyerabend who are better known to philosophers of science [3,4]. Popper considered Kant's problem of *demarcation* to be the central issue of science, i.e. how we demarcate scientific theory (or more simply, science statements) from those of metaphysics (Kant's concern) or of pseudo-science.

Popper was highly influential in philosophical circles for some time, although his work is of less current interest to philosophers of science [3,4]. However, he has had a lasting influence on scientists and engineers and on the issue of Validation of computational physics models.

(A note for technical readers only: by *computational physics models* we are concerned with models based on the approximate numerical solution of nonlinear partial

A Defense of Computational Physics

differential equations, the solution of which requires use of modern scientific computers and equally advanced algorithms. However, the philosophical arguments would be applicable to simpler models as well.)

It is my contention that, although Popper's work [1] is well-known, it is in fact not well-*read*, and that if it were well-read it would be less respected as relevant to computational physics modeling. One assumption by Popper [1] that immediately strikes a non-philosophy specialist is that science hypotheses should have the form "All x are y." This formulation does not inspire recognition in modelers! The following is a shorthand version of Popper's philosophy of demarcation (also known as non-verifiability or *falsificationism*) as usually presented.

2.2 *Popular Popper Précis*

The following is my impression of the popularly held shorthand statement of Popper's theory of *falsificationism*.

> **Popular Popper Précis:** A scientific theory, and by extension a computational model, cannot be validated, i.e. proven to be true. It can only be invalidated or falsified, i.e. proven to be false. It must be capable (in principle) of being invalidated, i.e. be falsifiable, otherwise it is not a scientific theory but only a pseudo-scientific theory (or perhaps metaphysics).

To avoid the impression of setting up straw men, I give five examples of the use of Popper's assertions in significant *computational* works in modern context. Popper has been quoted as an authoritative witness to the fundamental impossibility of Validation of computational physics models by a National Science Foundation Blue Ribbon Panel on Simulation-Based Engineering Science [5, p. 34]. (This panel did indeed represent enormous experience and expertise in computational physics modeling.) Popper's philosophy was foundational to the widely cited papers by Oreskes et al. [6] and Konikow and Bredehoeft [7], the latter entitled "Groundwater Models Cannot be Validated." Oreskes et al. [6] claimed to limit their rejection of Validation (or *verification* - see below) to "natural systems" (in particular, proposed underground nuclear waste disposal sites) while accepting the possibility for manufactured systems. But this distinction is not so sharp, e.g. bridges are built without knowing everything about the bridge's supporting geology, aircraft fly in a "natural" variable environment, etc. In my own professional experience, these two papers [6,7] have been taken much too seriously and have caused real problems. More recently, Oden et al. [8] stated that "in line with Popper's principle, a model can never actually be validated." However, these authors counterbalance this deferential citation with practical and insightful observations on the true utility of models, noting that "the validation issues posed here are somewhat different from the usual falsification of scientific theories." Hazelrigg [8] gave less balanced pessimistic claims, invoking not only Popper but also Arrow's Impossibility theorem.

This importance in the computational physics modeling community is remarkable, considering that the applicable philosophical arguments appear in the first edition of Popper's most cited book, *The Logic of Scientific Discovery* [1], the first edition of which was published (in German) in 1934, well before the advent of modern computers and computational modeling. Whatever Popper's contributions or claims were, they were not directed specifically towards Validation of computational physics models, but to scientific theories in general.

2.3 Objectives

To be frank, I would like to drive a stake through the heart of *falsificationism* as applied to computational physics modeling and silence it once and for all. Though I doubt this is possible, I write this down-to-earth critique not in the hope of changing any true believer's mind, which would be highly unlikely, but to give some support to anyone who needs to address and dismiss the issue (because of reviewers, funders, regulatory agencies, stakeholders, etc.) and to be able to get on with their work.

I intend herein to critique Popper's philosophy, relying on the many criticisms already in the philosophy of science literature, plus a few perhaps original observations. The critiques will be given at three levels:

(A) **philosophy of science,**

(B) **empirical data on how science is actually conducted in the 21st century,**

(C) **applicability to computational physics modeling and the question of Validation.**

2.4 Critique Level (A). Problems for the Philosophy of Science

Prior to our main interest area of computational physics, Popper's philosophy of *falsificationism* has plenty of problems when applied, as he intended, to science in general.

2.4.1 Popper Invented *Falsificationism*, not Falsifiability

As a preamble and matter of historical credit, Popper did not quite invent the concept of falsifiability, although he emphasized it to the exclusion of all else. A reader could be excused for believing that Popper believed he did invent it, as he seemed to infer so in several places. Even philosophers of science slip into this claim, e.g. [3,10,11]. But Popper himself (on p. 17 of [1]) indicated that his self-proclaimed adversaries, the logical positivists (and later incarnations as logical empiricists and neopositivists) already used falsifiability as *a* criterion, but not as *the* criterion, as follows.

"The criterion of demarcation inherent in deductive logic ... must be capable of being finally decided, with respect to their truth and falsity; we shall say that they must be '*conclusively decidable.*' This means that their form must be such that *to verify them and to falsify them must both be logically possible.*"

Popper cites the logical positivists Schlick and Waismann of the Vienna Circle (Schlick had founded it) as sources of these statements. He disagrees with the logical positivists for applying both these criteria of deductive logic to the demarcation problem.

So Popper did not quite invent falsifiability, nor did he substitute falsifiability for verifiability. Rather, he examined the two criteria already proposed by the logical positivists in their description of "*conclusively decidable*" (or elsewhere "*completely decidable*"); then he rejected verifiability but accepted falsifiability. (I would say he *adopted* falsifiability, as in claiming responsibility for it.)

However, there is no doubt that Popper deserves the credit for developing the theory of falsifiability, for pursuing the ramifications, and for proselytizing the concept. S. O. Hansson[C] has pointed out to me that the logical positivists had emphasized verifiability over falsifiability, so Popper's contribution may be even stronger than I could infer from reading his works [1,2]. Popper certainly proselytized for falsifiability and became identified with it. Falsifiability would not have figured so prominently in 20th century philosophy of science without the work of Sir Karl Popper. It is clear that he deserves the credit for inventing *falsificationism*.

2.4.2 Demarcation

Popper's concern was with the "problem of demarcation," separating science from pseudo-science based on the single, simple criterion: a true scientific theory is falsifiable, whereas pseudo-science theory is not. Strictly, the problem of demarcation of science from pseudo-science is different from the demarcation of science from metaphysics [10,11], but Popper thought falsifiability worked for both. A particular field of his interest was Freudian psychology, which Popper considered not science, with its unfalsifiable constructs like Id, Oedipus complex, Superego, etc. Two other fields of interest which he found at least initially plausible and appealing [2, Ch. 1] but which failed his demarcation criterion were Adler's "individual psychology" and Marx's theory of history. Astrology was the obvious archetype of pseudo-science. Popper's idea of a supreme scientific theory was Einstein's relativity, with its startling and revolutionary re-conception of the universe, and importantly with its predictions which in principle are capable of being be falsified. According to Hansson [11], in Popper's last statement of his position he left no doubt that he considered falsifiability to be the only criterion, both necessary and sufficient.

"A sentence (or a theory) is empirical-scientific if and only if it is falsifiable."

This statement presents an extreme position which really cannot be supported. Not every statement that is incorrect deserves to be labeled pseudo-science; for the derogatory prefix *pseudo-* to be deserved, there must be non-science *posing* as science. This quality is not included in the demarcation criterion of falsifiability. The fundamental orientation of Popper (and many of his contemporary philosophers of science) was towards minimalist absolute (perhaps, *atomistic*) statements to settle vexing questions. The esthetic of economy is evident in their proffered candidate statements, but the economy is lost in the ensuing discussions, rebuttals, refutations, clarifications, and elaborations, all of which are typical of a legalistic *definition* rather than a *description*. In the end, initially wordy descriptions turn out to be more economical, as well as more understandable, helpful, realistic, and honest, in my opinion.

Hansson [11] cited 14 later philosophy papers that have offered *lists* of characteristics, usually 5-10 items, rather than a single criterion, to identify pseudo-science or better, pseudo-scientific practice. His own list of 7 items follows (see [11] for elaboration).

A Defense of Computational Physics

- Belief in authority
- Nonrepeatable experiments
- Handpicked examples
- Unwillingness to test
- Disregard of refuting information
- Built-in subterfuge
- Abandonment of explanations without replacement

This approach differs from that of Popper and his contemporaries in at least three ways: its rejection of minimalism, its descriptive rather than definitive character, and the *focus* of its demarcation. Popper and his contemporaries focused on *theories* (or more basically, statements) while the latter approach focuses on *behavior* of the community. This orientation began with Kuhn [12]. For example, alchemists of old should not be labeled pseudo-scientists. How were they to know that elements could not be transmuted? Their techniques were not always scientific by modern standards, but the honest ones paved the way for the modern science of chemistry. So instead of asking, for example, "Is Creationism a Pseudo-Science?" a better question might be "Do advocates of Creationism behave pseudo-scientifically?" Based on Hansson's list above, my answer is "yes" on 7 of 7 criteria. Many would believe this approach also retains, i.e. agrees with, Popper's decision on the pseudo-science status of Freud, Adler, Marx, and astrology as practiced during Popper's time. The unexpected observation that very different demarcation criteria often lead to the same rejections was observed by Kuhn [13] and explained recently by Hansson [14].

On the other hand, most scientists (including religious believers) believe in the theory of biological evolution (and more importantly, the facts of biological evolution); in our opinion, biological evolution is science, not pseudo-science.

Popper long disagreed, claiming that the theory of evolution reduced to the useless tautology "survivors survive," and only grudgingly came around to accepting it as a science, though uniquely difficult to categorize.[D]

2.4.3 Changing Terminology

There is an annoying problem today of inconsistent terminology. Popper's philosophy has been used to discredit claims of Validation of modern computational physics models, as noted [5-9], but the terminology has changed. Popper's cited work was first published in 1934, well before modern computational physics modeling existed. Although his book [1] was much revised (tortuously, with footnotes to footnotes) through 1980, Popper (in English translation) did not use the word *Validation* nor even *validate* (except in a few obscure footnotes) but rather *verify*. Yet his philosophy has been used to rebut claims (or to pre-emptively disallow future claims) of *Validation* of computational physics models. In modern usage, one speaks of the area of "V&V" for Verification and Validation.

Although there are several semantic disputes in modern V&V, the distinction between Verification and Validation is not one of them. "V&V" is really three subjects, not two, and the following description is widely accepted, notably in ten modern publications in V&V for computational science and engineering [15-24], including two ANSI Standards (American National Standards Institute). The three subjects are Verification of Codes, Verification of Calculations (or Solutions), and Validation.

Briefly, Verification of Codes concerns correctness of the coding, i.e. the computer code is free of significant mistakes and does what its documentation claims. If the code manual claims to compute space shuttle trajectories to (say) second-order (mathematical) accuracy using a flat earth Newtonian model, and it is proven (i.e. convincingly demonstrated) that the code indeed solves these equations to second-order accuracy, then the code is Verified. Verification of Calculations involves error and uncertainty estimation for a case of interest. (Note that the second Verification assumes the first has been done; a Verified Code can be used for many different calculations.) Both Verifications have to do only with mathematics. There is no question, for example, of whether or not the flat earth equations provide adequate agreement with the physical reality; that question is the subject of Validation. Only during Validation is the accuracy of the physics evaluated.

The suggestion of meaning from etymology is sometimes invoked [5-7]. Although etymology cannot override what people collectively decide a word means, especially for technical terms, in these present cases it suggests meanings close to accepted use in the technical area. Verification (from the Latin root) suggests truth, a good connotation for Verification of Codes and perhaps appropriate though weaker for Verification of Calculations. Likewise, the (Old English) root for Validation suggests strength of argument or position, which also is appropriate, or at least has seemed so to several authors. Some authors (including [6,7]) place much importance on the etymology, but in early days of computational physics there was not uniform usage. The conceptual distinctions between Verifications and Validation are important, but the particular choice of terms is not (as Popper would probably have agreed); it is, however, a widely accepted and therefore convenient convention. It does not agree with Popper's terms (as translated from the German), so some tedious care is required for discussion.

2.4.4 Non-Computational Difficulties with Popper's *Falsificationism*

Philosophers of science have pointed out other substantive difficulties with Popper's philosophy, aside from our concerns here with application to Validation in computational physics modeling.

A problem common to all philosophy of science is statistics. Science is not as black-and-white as it used to be in the good old days. Inherently variable systems (and placebo effects) mean most scientific theories are only testable statistically. Widespread use of "significance" (meaning statistical significance, not meaning *large* but possibly small and probably real but not completely sure ....) makes the subject fuzzy at best. "Statements involving probabilities appear unfalsifiable" [4].

Popper's demarcation philosophy may be used to identify (say) Creationism as pseudo-science, since the claims are not falsifiable in principle, as opposed to relativity theory, his favorite example. The trouble is, we would like to use the term *science* for more than physics and its cousins. There are real problems defining falsifiable statements in social sciences (and there are serious objections to claiming these as *science*), and there are other problematic fields as well. Linguistics, for example, has branches that are certainly rigorous and non-debatable but involve nothing more than observation and taxonomy building. But when linguistic *theories* are attempted, the falsifiability is not evident, e.g. Chomsky's theory of meta-language (now of little interest) or theories on origins of language, with unfalsifiable assertions, and indeed with lack of consistent definition of *language*. Cosmology, the favorite popular science, is so weakly supported (i.e. constrained) by observational data that one can safely bet that popular science magazines will announce findings that are expected to "revolutionize" basic conceptions of our universe on a regular (almost annual) basis. This borderline pseudo-science position for modern cosmology has been asserted by some physicists and cosmologists [25-27]).

Again, this demarcation is corrupted by statistics. The exciting and extremely useful advance of modern genetic dating rests on assumptions (we would say, modeling assumptions) of nominally constant rates of change, i.e. mutations. Even then, the uncertainty statements can be large, as in "....[scientists] estimate that the mitochondrial ancestress of humans and Neanderthals lived 465,000 years ago, give or take a couple hundred thousand years either way" [28]. Such a huge range of uncertainty does not completely negate the usefulness of the estimate (although there is a natural tendency to carry around and use the best estimate while forgetting the uncertainty range) but it surely does complicate the issue of falsifiability. If new independent data were to fix a date of 685,000 ± 30,000 years, would this qualify as a falsification of the original statement, even though the uncertainty range (error bars) for the two estimates overlap (by a range of only 10,000 years)? Or would this constitute Validation? Very weakly, if at all. Then, would it perhaps constitute Invalidation, i.e. an analog to Popper's falsification? Arguably "yes", but only probably. Yet it is science, most would agree.

A prime candidate for the successful application of falsifiability to the demarcation problem is astrology, but it is not so simple. As pointed out by Hansson [11], Kasser [4], Godfrey-Smith [3] and others, if astrology is pseudo-science, then its claims cannot be falsified. But in fact it now has been tested, and initial tests famously showed a small but statistically significant correlation between personality traits and astrological predictions. Later the tests were refined and it was recognized that the personality traits, which importantly were self-reported, correlated somewhat for those study participants who believed in astrology, but not for those who did not believe. All but "true believers" conclude that astrology has been falsified. This story is all unsurprising and of little interest to those of us who consider astrology silly to begin with, but as Kasser [4] points out, it does put a statistical cloud over the fact of falsification.^{E} Could not a believer claim that astrology is not pseudo-science but rather just a real scientific area that needs more work to solidify its claims statistically (e.g., like much of modern pharmacology)?

Further and most importantly, astrology debunkers cannot, with logic, claim that astrology is pseudo-science because its claims are not falsifiable, and simultaneously claim that those claims have been falsified! Which is it, pseudo-science or just not yet very good science, whose conceivably weak yet statistically significant correlations have yet to be nailed down? The debunkers (myself included, I hasten to clarify) cannot have it both ways. The same could perhaps be said for at least some of Freud's models, and other areas that many of us would like to dismiss as pseudo-science. Kasser [4] points out that the ongoing debates over the teaching or exclusion of Creationism (now in its later incarnation as Intelligent Design) have been fought in courts on the basis of claims and counter-claims of pseudo-science, but a simpler and more direct challenge might be to reject it on the basis of being, at *best*, poor science and not worthy of being taught, like phlogiston theory.

If we accept (as I certainly do) that astrology now (post-Popper) has been conclusively falsified, then we would have to say (1) the tenets of astrology are wrong, but (2) following Popper's demarcation, astrology is not pseudo-science, i.e. its statements (theories) are falsifiable and hence it should not be categorized as pseudo-science by Popper's demarcation criterion of falsifiability. Popper then would have to eat his own words because he considered this failure to constitute a damning evaluation of Wittgenstein's proposal for demarcation [2, p. 53]. Popper claimed Wittgenstein's demarcation could not identify astrology as pseudo-science, and this was evidence that it could not be taken seriously as a demarcation criterion.

2.5 Critique Level (B). Empirical Data on Science Practice: *Falsificationism* Falsified

The pre-requirement of Popper's philosophy, before *falsificationism*, is *empiricism*, shared with his self-proclaimed adversaries the logical positivists. Any statement aspiring to scientific status (or more generally, having "epistemic warranty" [14]), should be tested empirically. S. O. Hansson [10] has done just that, testing Popper's philosophy to determine if it is actually used by scientists. (Hansson is eminently qualified for this work.[F]) His brief conclusion is in his title: "*Falsificationism* Falsified."

Hansson performed a conceptual analysis of Popper's *falsificationism*, dividing the central thesis into several components. Popper's assumption [1] is that all science theories to be tested (possibly falsified) can be cast in the form of a hypothesis with a simple binary (yes/no) outcome. "Popper's conviction that hypotheses can be falsified but not verified seems to have been based on the view that hypotheses should have the form 'All x are y'. Statements of this form can be falsified by a single observation." It requires only little experience of doing or reading science to recognize that this in fact is not a common formulation of science practice. So the most fundamental condition to ascertain is whether or not there is a hypothesis to be tested!

As Hansson noted, this condition eliminates explorative research, described as that which "proceeds open-endedly without any preconceived categorization of possible research outcomes." It also eliminates research whose results could cover multiple (> 2) interpretive categories, and others. The only category amenable to Popper's formulation is the following (see [10] for exposition).

(a) *predetermined* (i.e., not explorative) **research with**
(b) *binary interpretive structure* **(yes/no outcome), possessing**
(c) *status asymmetry* **which is**
(d) *confirmatory* **and**
(e) *negatively coupled*.

Hansson then took an empiricist stance and actually *looked* at what scientists were doing. For his sample, he chose the high-status, broad scientific coverage of the prestigious journal *Nature*. He analyzed all 70 articles appearing in *Nature* in the year 2000. Some highlights of his results follow.

Hansson found that 49 of the 70 articles were explorative in the sense that, as far as can be seen from the text, the authors had not identified the possible outcomes beforehand and divided them into distinct categories. One of the remaining 21 articles had done so but used three categories. The remaining 20 of 70 articles had a binary interpretative structure. Three of these 20 had no "status asymmetry," meaning that the two outcome categories were treated equally, so that neither could be considered a hypothesis. In 8 of the remaining 17 articles, no clear confirmatory asymmetry was found. See [10] for details. Two particularly interesting articles revealed positive confirmatory asymmetry that, contrary to Popper's prescription, were "much more conclusive than what a negative finding could have been."[G]

For a bottom line evaluation, consider the following. Only 2 of the 70 articles tested a hypothesis that was more accessible to falsification than to verification (our Validation). Only one of these had a negative outcome, i.e. "conformed to the falsificationist recipe."[H] (Hansson noted that his structural classification of being more accessible to falsification than to verification involves some judgments on his part; ignoring this part of the classification, it remains that in only 3 of the 70 articles was a hypothesis falsified.)

As Hansson noted, Popper's model of a hypothesis statement in the form of "All x are y" is not often met. Judging by his study of *Nature* articles (and confirmed in common experience), "most modern high-status research is either exploratory or devoted to hypotheses that do not have this logical form." One could argue that the structures of these research projects *could* or *should* have been so reformulated, but it is not evident how this could be done in a way that "conforms both with the tenets of *falsificationism* and with the state of knowledge in the respective field." For such explorative research "no hypothesis could have been posited in a non-arbitrary way" [10].

What is a falsificationist to do in the face of this empirical evidence? The remaining option is to deny the quality of the research. (Might a true believer falsificationist perhaps even dare to label 69 of 70 articles in *Nature* as *pseudo-science*?) Hansson states that his study confirms the view of Nagel [29] that

> Popper's "conception of the role of falsification in the use and development of theories is an

oversimplification that is close to being a caricature of scientific procedure."

Hansson [10] concludes as follows. "In other words, *falsificationism*, as a methodological recommendation, is of very limited value since the conditions for its application are not in general satisfied. *Falsificationism* presupposes that good research has a structure to which the falsificationist strategy is applicable. The results reported here constitute a falsification of that presupposition. This falsification extends to *falsificationism* itself, in the sense that it should be subject to the same treatment that it itself recommends for falsified hypotheses."

Briefly, *falsificationism* has been *falsified* empirically.

2.6 Critique Level (C). Popper's Philosophy Applied to Validation in Computational Physics Modeling

For the sake of argument, let us imagine that, on a rarefied philosophical level, we accept Popper's philosophy, that only falsifiability reigns and that there is no chance of verifiability of any scientific theory. Let us further ignore Hansson's [10] conclusive demonstration, corroborating the intuition of practicing scientists and engineers, that Popper's prescription of how science is done or should be done is in fact not representative of good science practice. Still a question would remain: Is Popper's *falsificationism* as usually stated applicable to Validation in modern computational physics modeling? My conclusion, and that of some other modelers who have thought about it, is that it is not applicable.

A Defense of Computational Physics

These further arguments follow in several sub-levels.

2.6.1 Critique Level (C-1). Common Sense

My own biggest problem with Popper's philosophy is simply, in technical terms, its airy-fairy-ness. If this evaluation of Popper sounds too harsh, consider his claim, [1, p. 444].

> **"The statement 'this container contains water' is a testable but non-verifiable hypothesis, transcending experience."**

And, I would add, transcending common sense. I can tolerate (barely) Popper's position, given his perspective and intellectual tradition, but I cannot understand how practical scientists and engineers could take this stuff seriously. A possible reason is that they have not read it, except in modulated second- or third-hand forms.

There is an even more serious affront to common sense by Popper. He also claimed, in accord with Hume, that there is no logical reason to prefer an extensively tested theory, say for a construction project, to an untested theory. The most he could align with common sense was to allow that it is *no less* logical to use a tested theory than an untested theory. I recognize that Popper's was a great mind, but really, this position and this kind of thinking are simply impossible to take seriously.

Perhaps it is true that we, as a technological culture, cannot adequately model underground nuclear waste disposal methods, or climate changes, but such evaluation should not hinge on a philosophy of science that equally leads to the conclusion that we cannot say with confidence that a cup contains water, and that there is no "logical" reason to prefer a tested theory to an untested one.

2.6.2 Critique Level (C-2). Are Models Equivalent to Theories?

Popper was concerned with the evaluation of a scientific theory, whereas computational scientists and engineers are concerned with the evaluation of a computational model. Is *theory* equivalent to *computational model*? Mehta thought so [30; also cited in 15,16], and likened the task of Validating a computational physics model to the general problem of Validating scientific theories. There certainly is similarity, but not everyone sees it as decisive. Some authors state otherwise, and note that the conceptual model behind the computational model is more analogous to scientific theory. (E.g, see Rykiel [31,32], Refsgaard and Henriksen [33].)

Rykiel [32] also noted that a mathematical model is neither a hypothesis nor a theory, and that *un*validated models, though not dependably and quantitatively predictive of nature, can nevertheless be useful in a scientific endeavor, and often are useful in ecological system modeling. See also discussion by myself in [16, Section 9.2.3]. Rykiel also wrote, "These terms [including Validation of computational physics models] are defined in a limited technical sense applicable to the evaluation of simulation models, and not as general philosophical concepts. ... Single minded focus on falsification is a superficial treatment of a complex subject..." He noted, with examples, that "The impulse to falsify can result in 'naive falsification'."

But again for the sake of multi-level arguments, we can tentatively accept the parallel between science theories and computational physics models. Then the Level (C) Critique can be further subdivided.

2.6.3 Critique Level (C-3). Inclusion of Popper's *Numerical Universality* in Validation

Proponents of "impossibility of Validation" attribute their position to Popper's *falsificationism* but do not account for Popper's distinction between *strict universality* and *numerical universality*. The latter *is verifiable* according to Popper himself, and I claim it is the appropriate association for modern Validation of computational physics models. It is appropriate not merely because of Popper's use of the word *numerical*, which might seem to suggest *computational* but is actually used differently (see discussion by Popper [1], pg. 40). Today it would be better termed simply *finite*.

In current normative V&V practice for computational physics modeling [15-24], Validation is *not* claimed for a continuum of experimental set points (parameters) but only for a finite number of experimental conditions. If a good model Validation agrees with experiments at all the set points tested, then it qualifies for Popper's description of *numerical universality*, i.e. *universal* because it agrees at *all* the existing experimental set points. Any attempted extension of Validation results towards Popper's concept of *strict universality*, implying a continuum (uncountable number of cases), arguably is not part of Validation *per se*. This position is not quite universally agreed upon, though it is the position of [15-24]. In this view, such extension is part of practical use of a computational physics model (e.g. in decision making for a project) and is usually expressed as legitimacy of interpolation, producing a continuum of cases (sometimes described as a *domain of validation* [15-24] but better as a *domain of application*).

However, it is a defensible position that extension to a continuum domain of Validation should properly be considered part of Validation. But the more modest claim of Validation only at experimental set-points (i.e. the specific finite number of conditions) constitutes a practical and advantageous division of labor, because there is only slight possibility of claiming accuracy for interpolation of Validation results without making unjustified mathematical assumptions.[I] In most cases, experimental data and therefore Validation assessment comes in discrete packages, and in current normative practice [15-24] Validation is claimed only at these discrete set-points, finite in number. In any case, everyone agrees that extension to a continuum Domain of Application is necessary for the results to be useful in applications away from the experimental set-points.[J]

2.6.4 Critique Level (C-4). Popper's *Truth* vs. Validation *Accuracy*

The key issue of this booklet is: Does *Validation* of a computational physics model imply *Truth* of the model?

As noted previously, Popper did not use the terms *Validation* or *Validate* but rather the term *verify*. But critics of computational physics apply his criterion to concepts described by the modern term *Validation*, which involves comparison of computational answers with physical reality. (See Chapter 2.) I have argued against the domination of etymology, but I must admit that here the effect is telling. As suggested by his use of *verify*, Popper was indeed concerned with the *Truth* of a scientific theory.

Actually, *Truth* is highly problematic to philosophers. From Glanzberg's essay [34] "on the main themes" of *Truth*:

A Defense of Computational Physics

> "Truth is one of the central subjects in philosophy. It is also one of the largest. Truth has been a topic of discussion in its own right for thousands of years. Moreover, a huge variety of issues in philosophy relate to truth, either by relying on theses about truth, or implying theses about truth. It would be impossible to survey all there is to say about truth in any coherent way."

Glanzberg's coverage of just the "main themes" of *Truth* used 13280 words and 85 references.

The *Truth* of a scientific theory was Popper's concern. But in the modern use of *Validation* in conjunction with computational physics models, we are not concerned with some convoluted, vaguely or tortuously defined concept of *Truth* but rather with the simple, well-defined concept of *accuracy*.

In the testing of scientific theory (at least in the classical, pure sense) the question is one of correctness of the model, in the sense of *Truth*, whatever that may mean. But in most computational modeling (and generally so in engineering modeling) the question is not one of correctness or *Truth* but of mere accuracy, and whether or not the accuracy is sufficient for possible intended purposes.[K] And by *accuracy* in this context we refer to simple approximate agreement between the answers from the computation and the experimental values. We can Validate a dozen different RANS turbulence models for various intended uses even though they do not precisely agree with one another[L]; in a purist sense of (Popper's) *verification* of a theory, at most only one could be *True*. But our situation is even worse, in that we know *a priori* that a RANS turbulence model will not be *True* because of the method in which it was derived, i.e. the averaging process of deriving the "Reynolds Averaged Navier-Stokes" equations involves an approximation.

I have used the caveat "at least in the classical, pure sense" because I do not accept this attribute of exclusivity or uniqueness of *Truth* even for scientific theory. We may fantasize an alternative history of science in which Einstein and Newton switched places. Suppose that the equations of relativity were developed first, and after a few centuries, Newton invented the approximation to develop his equations. Would his theory be only grudgingly allowed for mundane engineering applications, or would it be recognized as a major scientific advance? I think the latter, and would even go so far as to claim *Truth* for the approximate equations. This theory really describes how our world works, at virtually all scales normally accessible to humans. Other examples abound.[M] See also Kasser [4] for the view that *less* complete theories can lead to *more* understanding and can be considered advances. In a similar vein, the usual distinction between engineering analysis and science, although somewhat meaningful, is not as sharp as often stated.

It is ironic that both Popper and his adversaries, the logical positivists, had such an aversion to metaphysics, yet they were preoccupied with the question of *Truth* of scientific theories. Many philosophers consider the concept of *Truth* to be essentially metaphysical [34], and Popper's use echoes this approach, in my opinion. On the other hand, philosophers can also get cautious over the word *accuracy*, sometimes taking it to refer to the essential correctness of a theory, which sounds more like *Truth*. They sometimes prefer words like *coherence* or *congruence* or *correspondence* or *empirical adequacy* for what scientists

A Defense of Computational Physics

and engineers (and the public) think of as *accuracy*. Here, we take the normal, work-a-day position that *accuracy* of a computation refers to the degree of agreement between the outcome of the computation and the correct value. (Also known as [18,19] the true value - there's that word *true* again.) In different contexts, *accuracy* can refer either to the (mere) mathematical agreement between a computation and the exact solution of the modeled continuum equations (Calculation Verification), or to the degree of agreement between the outcome of a computation and physical reality (Validation).

Many (not all) scientists would agree with Oreskes [35, p. 314, 368] that Newton's laws of motion and Einstein's theory of relativity cannot both be true.[N] Yet all would agree that both theories can be demonstrated to be *accurate* within well-defined parameter ranges, i.e. a *domain of validation* in our computational physics terms. In fact, the Newtonian laws are as accurate as Einstein's for any human-scale problem and tolerance, and are much preferable because of their relative simplicity. Popper would have argued that neither could be claimed to be *True* or verified (our Validated) but he allowed the word *corroborated*. Oreskes et al. [6] prefer that the term *confirmation* should be applied to what everyone else calls Validation of computational physics models, on the basis that *confirmation* is less misleading to the general public. But as noted by Rykiel [32] and myself [15, Appendix C], all three terms - verification, validation, confirmation - are synonyms in general non-technical use, so one is no more misleading than another.[O] The solution is not to depend on connotation but on explicit *technical* definition (Chapter 3).

The question "Does Validation of a computational physics model imply *Truth* of the model?" can be difficult. *Truth* can be binary, yes or no; I did take the cookies or I did not. But beyond such simple binary questions, *Truth* is not so well defined. Is it *true* that the earth is flat? No. Is it *true* to use flat earth assumptions and equations to model a baseball pitch? Perhaps "No." Is a computational physics model that uses constant g and flat earth geometry to model a baseball pitch *not true*, i.e. *wrong* or *false*? What about the use of Newtonian mechanics (g and $F = ma$ with constant mass m) without a relativity correction? Is that *false*?

It is much more profitable to change the question and avoid the question of *Truth*. The better question is, Is the computational physics model *accurate*? Easy question, unambiguous answer. Flat earth and Newtonian mechanics are so accurate for a baseball pitch that no one would believe any experimental contradiction. Of course, the baseball pitch would be extremely difficult to model computationally, but the discrepancy lies not with *UnTruth* of flat earth and Newtonian mechanics but with impossibility of accurately measuring initial conditions, characterizing the surface, turbulence modeling, etc.

Validation is used by (most) practicing engineers and scientists concerned not with "the whole truth about a phenomenon" but with the modest and realistic goal of *accuracy* defined for specific metrics and parameter domains. Whereas the concept of *Truth* for non-binary questions is fuzzy, *accuracy* can be well defined. The questioner can define accuracy in terms of % deviation from physical reality (measurements) or from a better, higher order theory (e.g. relativity). Then the responder can state the measured accuracy as (typically) some % deviation with another % uncertainty (or error bar), as in "Newtonian mechanics and flat earth agree accurately for this problem to within x % with an uncertainty estimate of y%." If the questioner sets the question in terms of an accuracy level and associated uncertainty, i.e. "Is the flat earth and Newtonian mechanics model accurate to within x % with an uncertainty estimate of y%" then the responder can answer *yes* or *no*.

When we give up the misguided quest for *Truth* and settle for the more modest, testable, meaningful, attainable, well-defined, non-metaphysical, and useful goal of *accuracy*, the operational consequences are significant. There is no longer any concern about deciding the one (possibly) *True* theory or model. There is no conceptual roadblock to Validation of multiple models which are logically mutually exclusive from the viewpoint of *Truth*. As previously noted, the prime example from CFD is provided by the many (maybe 10-20 serious) RANS turbulence models, none of which aspires to *Truth* or agrees fundamentally with the benchmark model of the full Navier-Stokes equations, but all of which can legitimately claim some useful accuracy in some metrics.

Truth questions actually can sometimes be appropriate in computational physics modeling. For example, during backfitting of groundwater flow models, the modeler might posit the existence and position of an unseen fracture, or perhaps a non-porous block of material. This modeling feature can be susceptible to a *yes/no* truth test; the feature is there, or it is not. But other modeling evaluations - in my experience, the overwhelming majority - are not of this type, but are simply accuracy issues.

This preference for *accuracy* over *Truth* might seem to echo the philosophy of *instrumentalism* in science theory, but in fact it does not preclude a non-instrumentalist higher regard for theory. Like most, we believe that successful theories do embody truths, and we seek more than mere practical predictive ability. In fact we see *Truth* where Popper refuses to acknowledge it, in Einstein's relativity and even in Newton's mechanics. Consider the following quote from Albert Einstein himself.

> "There could be no fairer destiny for any ... theory than that it should point the way to a more comprehensive theory in which it lives on, as a limiting case."

This evaluation hardly suggests that Einstein considered Newton's theories to be *false* but only that they were superseded, only in extreme parameter ranges, by his more complete theory. Curiously, this quote was used approvingly by Popper himself, to introduce his first chapter in [2].

We can reject mere *instrumentalism* for our science theories while still modestly claiming only assessment of accuracy by the term *Validation* for computational physics models. Whether the underlying conceptual/mathematical model or theory behind the computational model represents *Truth* or not, or whether it so much as aspires to *Truth* in some sense, is a separate question and is fair game for all Popperian and non-Popperian adversaries; in any case [36] (see Chapter 3), what we mean by *Validation addresses only accuracy*.

The practical issue is, "accuracy of what?" The answer is, accuracy of the Validation quantities.[P] The questioner decides these, usually chosen with applications in mind. It may be surprising to find that Validation of computational physics models can and has been defined in two ways with regard to whether or not a pass/fail criterion is included, i.e. whether or not the Validation accuracy as determined is adequate for a candidate application. See discussion, including history and recommendations and repercussions, in Chapter 3. In fact, most V&V specialists (including myself, now) argue that it is preferable to *not* include a pass/fail tolerance, or acceptable level of accuracy, in the definition of Validation [16,19-21]. Rather, the results of a Validation exercise are simply reported as a level of accuracy and an uncertainty estimate, leaving the specification of acceptable levels to another part of an engineering project, usually described as *Certification* or *Accreditation*. The same level of accuracy may be acceptable for one application but unacceptable for another. Really terrible models, e.g. those that do not even predict correct trends, can be decisively "invalidated" for some parameter range, but more often one can neither determine universal criteria for acceptable accuracy nor for unacceptable accuracy, i.e. universal criteria for Validation or InValidation.

However, the questioner or model user should specify the quantities to be evaluated for accuracy. This specification is critical to understanding the division of labor in the application of Validation results, i.e. in the usefulness of a computational physics model. It is the critical issue in the following example.

2.7 A Provocative Example of Validation of a Computational Model

As a provocative example of what we modestly mean by "Validation of a computational physics model," which may be quoted out of context to the considerable amusement of one's audience, let us consider the question of Validation of a well-known astronomical model used for computations, the predictive accuracy of which is already well established. In fact, historically it has proven its accuracy arguably longer than any other scientific theory or model. I refer, or course, to the Ptolemaic theory of the geocentric motion of planets, used and trusted and, I would argue, *Validated*, for well over a millennium [37].

Ptolemy of Alexandria (~ 100-170 CE) was a geometer and is widely acknowledged as the greatest of ancient astronomers. He described his theory in his *Syntaxis* (later *Almagest* in Arabic). He was also an astrologer, of course, writing *Tetrabiblos* [38].

As a computational model, Ptolemaic theory admittedly is a bit complicated, based as it is on an earth-centered rotation of the planets. The diagrams are amusing, with little gear-like rotations and counter-rotations (smaller circular epicycles moving along larger circular deferents) introduced to account for observed features such as variations in observed brightness of planets and retrograde motion, e.g. the observed eastward path of Venus loops around and appears to move westward for a month or so (near inferior conjunction). For an engaging animation of a Ptolemaic system, see the following website.
http://astro.unl.edu/naap/ssm/animations/ptolemaic.swf

As a scientific theory, The Ptolemaic system was finally supplanted by the Copernican view of sun-centered rotations, demonstrated by Galileo in 1609 (published 1610). So the Ptolemaic model is "not *True*." However, the model users (including astronomer/astrologists who had very practical motivations as well as mystical ones) continued to use it for another hundred years, even past Newton, because it was easier to use for computations than the elliptical heliocentric orbits of Newton. As a computational model, it was *accurate*, i.e. in our terminology, Validated. The following description and evaluation is from Chaisson and McMillan [37].

A Defense of Computational Physics

> "By tinkering with the relative sizes of epicycle and deferent, with the planet's speed on the epicycle, and with the epicycles' speed along the deferent, early astronomers were able to bring this "epicyclic" motion into fairly good agreement with the observed paths of the planets in the sky. Moreover, the model had good predictive power, at least to the accuracy of observations at that time."

The earliest recorded geocentric models (which go back conceptually to Aristotle) were not too convoluted, but improvement in prediction accuracy required refinement by *calibration*, in modern computational terms [15,16,21] or *tuning* (or less charitably by fudging, tinkering, or twiddling) in the model parameters. "Perhaps the most complete geocentric model of all time" [37] was developed by Ptolemy around 140 C.E. It required 80 distinct circles; not what one would call an elegant model, but accurate. It was used for local clocks, calendars, and navigation aids. By Ptolemy's model, the position of the sun and moon became "utterly predictable" and the "formerly strange and frightening aberrations" of eclipses were explained and foretold [Deakin, 38].

Galileo's demonstration of the *UnTruth* of the Ptolemaic model, enabled by the revolutionary use of his telescope, hinged on one critical observation: the existence of full phases of Venus. (His discovery of the moons of Jupiter and their motion about that planet also undermined the concept of geocentricity, but were more suggestive than conclusive.) These full phases of Venus are not discernable by the naked eye, but are seen in even the primitive telescope of Galileo. In the Ptolemaic model, only partial phases up to the "fat crescent" would be observable.[Q]

Significant to our illustration of Validation, the astronomer/astrologers who were the "users" of the computational model *did not care* about predicting the phases of Venus. This phenomenon, which became a variable of interest to Galileo, was not part of the "Validation metrics" or criteria defined by the users, or by the "intended use" in our modern V&V terminology [15-24,36]. Thus, based on the Validation criteria set by the computational model users, Galileo's observations were not much of interest. There are parallels in modern computational physics models, e.g. shear layer turbulence models that are useless for predicting optical propagation (and are therefore arguably "falsified") but are accepted (considered "Validated") for predicting quantities of interest to the CFD model users, such as viscous drag [15,16]. Also relevant to the evaluation of the Ptolemaic computational model compared to the heliocentric model is the fact that Copernicus had still clung to the idea of circular motions (because these embody Platonic perfection), and therefore to match the data he still required some epicyclic motions, though smaller and heliocentric in contrast to those of Ptolemy. "Thus, he [Copernicus] retained unnecessary complexity and actually gained little in predictive power over the geocentric model"[37].

There were other reasons for not rejecting Ptolemy so quickly, even using the criterion of *Truth* rather than mere Validation *accuracy*. (Contrary to current pseudo-scientific / historic myth, religious preference for geocentricism had absolutely nothing to do with it.[R]) The common sense arguments were compelling for many: we have no sense of motion, we are not spun off the earth, there is no constant wind in our faces, etc. On a more esoteric basis, the

Copernican model explained a new phenomenological observation by Galileo, but contradicted an older one. The geocentric Ptolemaic model, but not the heliocentric Copernican model, was consistent with the best observations of the day involving the lack of stellar parallax, which was predicted by the Copernican model but not observed. Direct evidence of the motion of earth was not given until 1728, and stellar parallax was not observed until 1838, two centuries after Galileo [37].

For a century or two, one could argue either way, or at least one could argue that the jury was still out on the *Truth/UnTruth* of the two models. One could not argue, however, that the Ptolemaic model had failed Validation in our sense of computational accuracy. Nor were later developments so obviously *True* at first encounter. Newton's laws required a very troubling concept, that of "action at a distance." The little gears of Ptolemaic theory, now laughable to modern readers, were later explained by Descartes as manifestations of vortices in the ether. (Not a bad idea, to this fluid dynamicist.) Descartes' vortices were debated until the mid-18th century, and the existence of ether was not dismissed until the 20th century.

In summary, the Ptolemaic theory is surely not *True* but the associated computational model remains Validated. It exemplifies our distinction between *Truth* and our modest claim of mere Validation.

2.8 Not All Models are "Wrong" !

Closely related to Popper's concept that science theories cannot be verified (our Validated) is an often repeated statement from George E. P. Box [41] originally intended for statistical models. "Essentially, all models are wrong, but some are useful." People use this quote or some variant, usually dropping the important qualifier "essentially" and generalizing to non-statistical models, stating "All models are wrong." It may sound worldly and sophisticated (and in the sense of *sophistry*, I would agree). But why pick on "models"? Why not all theories, and all generalizations? This statement is based (if it can be granted a serious base) on the assumption not only of *Truth* vs. accuracy, but *Truth* in an esoteric sense. Not all models are *wrong*, but I suppose all are *imperfect* at some scale.

Note the double standard here: Popper and followers conclude that we cannot prove a theory right but only wrong, but paraphrasers of Box know categorically that all models are wrong. If so, there would be no need for falsification.

The nonsense is easily avoided by recognizing first, that a *model* includes not only some core description (like the ideal gas law $PV = RT$) but also statements on its domain of application. So if one applies the formula to extreme conditions outside this domain (e.g. disassociation for $PV = RT$), one cannot blame failure on the model; the *formula* has failed, not the *model*. Second, one must recognize that a model is just that, a model, so the question to ask is not one of *Truth* in some strained sense, but of accuracy, which is not vulnerable to such nonsense.

2.9 Limited Domain of Validation

Another feature that demonstrates the opposition of Popper's concept of verifiability (which he rejected) and our Validation of a computational physics model is the domain over which it extends. Popper was emphatic in his admiration for Einstein's relativity theory, as it exhibits the characteristic of a great scientific theory: its extended applicability outside the original dataset that motivated the development of the theory. Though clearly an admirable accomplishment for a

A Defense of Computational Physics

scientific theory, this extension is *opposite* to accepted practice regarding the domain of Validation of a computational physics model.

No modeler could ethically claim that Validation of a model's prediction can be used far outside of its domain of Validation. Some extrapolation ("outside") is usually justified, as long as no physics boundaries are crossed during the extrapolation. [15,16,19-21] For high parameter dimension problems, the data set almost inevitably will be poorly supported and the distinction between interpolation ("inside") and extrapolation lessens in significance. However, the goal of a Validation program is to *avoid* the necessity for just the kind of wide ranging expansion of the domain of application that Popper admired so much. Any claim of using a Validated model far outside the support of its validating dataset would be unethical and would be rejected by peers, contract monitors, and regulators.

For example, a new turbulence model validated for simple conditions, say incompressible planar boundary-layer flows with no free surfaces, if it were applied to supersonic flow or swirling flow or free surface flow, would be considered a not yet Validated model. Popper would like best an un-validated model!

2.10 Popper and "Normal Science" of Kuhn

One of Kuhn's [12] most valuable insights is his recognition of the difference between "normal science" and what he terms "revolutionary science," with "crisis science" bridging the gap. Normal science proceeds within an existing paradigm, making progress without overturning any truly fundamental ideas of how the universe works. The archetypical example is the Newtonian world. There exists a significant mismatch in public perception between the importance of Einstein's theory in supplanting Newtonian mechanics and the reality of applied science. Before Einstein, there was much progress made in scientific discovery within the Newtonian paradigm, and that work and progress continued after Einstein, and still continues. Very few and isolated areas of science, and certainly of technology, depend on any correction of Newtonian mechanics from Einstein's theories. But in certain areas, e.g. astronomy and cosmology, Einstein created a revolutionary change of paradigm. Now, within this new paradigm, progress continues to be made in a "normal science" mode.

As previously discussed, Popper's philosophy of *falsificationism* of scientific theory is concerned with *Truth*. His view of science is highly ambitious, with what he considers the always unattainable goal of universality. He considers theory testing to be best if surprising results are predicted. We believe that his conceptual model of science is seriously flawed, but even *at best* it would be applicable only for Kuhn's [12] category of "crisis science."

Kasser [4] notes that Kuhn's normal science is not concerned with falsification; in fact, normal science tends to be dogmatic. Nevertheless, normal science is self-correcting; it is capable of undermining itself through illumination of inconsistencies. When this undermining happens enough, a crisis develops and the Popper crisis model, concerned with falsification, is appropriate and comes into play. Interestingly, in Kuhn's view this crisis is not so much logical as psychological, i.e. it is "felt." This description is realistic of what actually happens, but does not fit the emphasis on rationality by Popper and logical positivists. [4]

"Popper came to see the two standard virtues of scientific theories - explanatory power and confirmation by a large number of instances - as closer to being vices than

A Defense of Computational Physics

virtues" [4, #2]. "Fitting the data well is, thus, not the mark of a scientific theory; a good scientific theory should be informative, surprising, and in a certain sense, improbable." Does this orientation of Popper seem appropriate for a debate on Validation of computational physics models?

Popper's orientation presents a lopsided distortion of normal science, analogous to a discourse on political science that would be exclusively devoted to armed revolutions.

2.11 Contrasting Characteristics of Popper's *Falsificationism* vs. Computational Physics Model Validation

Here we reiterate some characteristics of Popper's theory of *falsificationism* contrasted to Validation of modern computational physics models.

Popper's *falsificationism* is concerned with *Truth*. His view of the scientific endeavor is highly ambitious, always aiming for universality and application beyond the realm of existing experiments, and valuing surprising results. At best, this view would be applicable only for Kuhn's category of "crisis science." Each of these aspects is opposed to computational physics modeling practice and accepted terminology. The practical Validation of a computational physics model is not concerned with *Truth* but simply with mere *accuracy*. This view is modest, with a usually attainable goal of practical accuracy. The surprising predictions that Popper values in science theory are *suspect* in computational physics.

Thus the overall process of Validation of computational physics models (necessarily preceded by Verifications of codes and individual calculations) is applicable not to the "crisis" and "revolutionary" science which Popper admired but to Kuhn's "normal science." Attempts to view the Validation of computational physics models from Popper's viewpoint of theory verification are not consistent with the normative concept of Validation within the computational physics community.

These contrasting characteristics are summarized in the following table.

Table 2.11.1. Some characteristics of Popper's concept of *falsificationism* of scientific theories contrasted with the accepted and normative concept of Validation of computational physics models.

Popper's *Falsificationism*	Computational Physics *Validation*
involves science theories	involves computational models
concerned with *Truth*	concerned with accuracy
ambitious	modest
surprising predictions are valued	surprising predictions are suspect
goal of universality	limited domain
Kuhn's "crisis science" or "revolutionary science"	Kuhn's "normal science"

A Defense of Computational Physics

2.12 Summary

Categorical claims, based on Popper's philosophy of *falsificationism*, of impossibility of Validation of computational physics models are not justified and can themselves lead to ethical difficulties. Popper's demarcation criterion of falsifiability is a valuable insight, but is not without philosophical problems even when applied to scientific theories, as he intended. It is not adequate for an "if and only if" demarcation of science vs. pseudo-science, as he had finally claimed. In "*Falsificationism* Falsified," Hansson [10] has empirically demonstrated that actual scientific practice in the year 2000 did not follow the Popper prescription in 69 of 70 cases examined.

When applied to Validation of computational physics models, Popper's philosophy of *falsificationism* as usually understood is inappropriate. However, Popper himself recognized a distinction that makes sense - that of "numerically exact" (finite in number) comparison which he did in fact recognize as "verifiable." For computational physics models, we would now say "Validatable." The inappropriate application to computational physics models of the dictum that a scientific theory is never verifiable but only falsifiable is not only incorrect but can cause much mischief, e.g. being used to categorically reject what possibly might be the best solution for nuclear waste disposal, or to categorically reject the possibility of usefully accurate climate modeling, etc. And the limitation of this dictum to "natural systems" [6] while allowing for manufactured systems is not such a sharp distinction, e.g. aircraft fly in a "natural" variable environment, etc.

Proponents of the impossibility of Validation of computational physics models often have a rarefied view of Validation that (a) has nothing to do with practical science and engineering and (b) is contradictory to widely accepted and pragmatic definitions of Validation as used by most computational physics modelers [15-24, 31-33, 36]. These include three graduate level reference monographs, two ANSI Standards, and three publication policy statements of scientific journals. These have a more authoritative claim to defining semantic distinctions and setting normative practice for computational physics modeling than citations of the philosopher Sir Karl Popper [1,2,5-9]. When there are genuine fundamental difficulties with computational physics simulations, as in groundwater flow modeling [e.g. 6,7], the difficulties will be related to obvious technical problems such as coarse mesh resolution, or lack of knowledge of physical properties, or inadequate accuracy of "laws" like Darcy flow, or uncertain model input such as rainfall, but nothing at all to do with Popper's *falsificationism* or Platonic ideals or *Truth*.

Of course, *falsifiability* is a tremendously important concept to science. But in the final analysis, Popper's philosophy of *falsificationism*, i.e. "falsifiability only" (a) is not defensible philosophically, (b) is not used significantly in (is not normative of) modern science practice, and (c) is neither applicable to modern computational physics modeling, nor endorsed by most of its practitioners.

2.13 Last Gasp

Finally, however one may judge the applicability of Popper's philosophy to computational physics models, it is certainly not less applicable to scientific theory in general, since that was his only concern. His position [1, p. 280] is as clear as it is useless to practical science and engineering; "every scientific statement must remain *tentative forever*." This statement is not to be confused with general scientific skepticism, nor with the universally accepted position of fallibilism. In rejecting Popper and this statement we are not claiming infallibility. Popper himself clarified that his claim "every scientific statement must remain tentative forever" involved *principle* and was more ambitious than mere fallibilism.

An anonymous reviewer stated the following observations. Normal scientific uncertainties and skepticism have been famously exploited by people with a vested interest in discrediting the science. Two older examples are biological evolution and health risks of smoking. That the same is currently true for climate models is not surprising [though it is still discouraging]. The response by the computational modeling communities must be to practice good computational science, and to be comforted by the fact that, despite the ever-present possibility of invalidation by future data, there are highly successful and reliable theories *and* computational physics models.

I completely agree with these observations. However, I must add that some of the many followers of Popper, even though intellectually honest and without vested interests, will not be satisfied with good practice of computational science. No matter how good a job we computational modelers do, *a priori* they will disallow the possibility of Validation of computational physics models as a matter of principle.

To avoid disputation and agonizing over what Popper or we may mean by *Truth*, we might grant his statement that "every scientific statement must remain *tentative forever*" in some rarefied and hopefully harmless sense, but note that Validation of computational physics models is thereby positioned in the same category as Newton's laws of motion and gravity, Einstein's theories, entropy, Darwinian evolution, conservation of mass, Fourier heat conduction, etc. We computational physics modelers are in good, respectable company.

Patrick J. Roache

Acknowledgements for Chapter 2

The author gratefully acknowledges discussions with Gabriel Arrillaga, Sven Ove Hansson, Jeffrey Kasser, William L. Oberkampf, Joseph Powers, Tina Riedinger, Kristin Shrader-Frechette, Thomas A. Stapleford, and Timothy G. Trucano, not all of whom agree with all my views. Also, two anonymous reviewers of a rejected journal article thoughtfully critiqued and improved my presentation. Thanks to Dr. Julie Ford for her professional editing. Patrick Knupp alerted me to two embarrassing errors of the first printing.

Some of this material is taken from my book [16], Section 9.2.2, "Objections to Validation Based on the Philosophy of Karl Popper."

References for Chapter 2

1. Popper, K. (1963), *The Logic of Scientific Discovery*, (translation of *Logik der Forschung*). Hutchinson, London, 1959. [Last copyright by 1980. Routledge version 2006.]
2. Popper, K. (1963), *Conjectures and Refutations: The Growth of Scientific Knowledge*. Routledge, London, 1963.
3. Godfrey-Smith, P. (2003), *Theory and Reality: an Introduction to the Philosophy of Science*, University of Chicago Press.
4. Kasser, J. (2006), *Philosophy of Science*, The Teaching Company.
5. NSF (2006), *Simulation-Based Engineering Science: Revolutionizing Engineering Science through Simulation*, Report of the NSF Blue Ribbon Panel on Simulation-Based Engineering Science.
6. Oreskes, N., Shrader-Frechette, K., and Belitz, K. (1994), "Verification, Validation, and Confirmation of Numerical Models in the Earth Sciences," *Science*, Vol. 263, No. 4, February 1994, pp. 641–646.
7. Konikow, L. F. and Bredehoeft, J. D. (1992), "Groundwater Models Cannot be Validated," *Advances in Water Resources*, Vol. 15, 1992, pp. 75–83.
8. Oden, T., Moser, R., and Ghattas, O. (2010), "Computer Predictions with Quantified Uncertainty, Part I", *SIAM News*, Vol. 43, No. 9, November 2010, "Part II" No. 10, December 2010.
9. Hazelrigg, G. A. (2003), "Thoughts on Model Validation for Engineering Design," DETC2003/DTM-48632, Proc. *ASME 2003 Design Engineering Technical Conferences and Computers and Information in Engineering Conferences*, Chicago, IL, U.S.A., 2-6 September 2003.
10. Hansson, S. O. (2006), "Falsificationism Falsified," *Foundations of Science*, Vol. 11, Springer, pp. 275-286. DOI 10.1007/s10699-004-5922-1.

11. Hansson, S. O. (2008), "Science and Pseudo-Science," *Stanford Encyclopedia of Philosophy*, http://plato.stanford.edu/pseudo-science.
12. Kuhn, T. (1962), *The Structure of Scientific Revolutions*, 2nd edition, enlarged, 1970; 3rd edition, 1996. University of Chicago Press, Chicago.
13. Kuhn, T. (1974), "Logic of Discovery or Psychology of Research?", in *The Philosophy of Karl Popper*, edited by P. A. Schilpp, pp. 798-819, La Salle, IL: Open Court.
14. Hansson, S. O. (2009), "Cutting the Gordian Knot of Demarcation," *International Studies in the Philosophy of Science*, Routledge, Vol. 23, No. 3, October 2009, pp. 237-243.
15. Roache, P. J. (1998), *Verification and Validation in Computational Science and Engineering,* Hermosa Publishers, Albuquerque, 1998.
16. Roache, P. J. (2009), *Fundamentals of Verification and Validation*, Hermosa Publishers, Albuquerque, 2009.
17. AIAA (1998), *Guide for the Verification and Validation of Computational Fluid Dynamics Simulations*, AIAA G-077-1998, American Institute of Aeronautics and Astronautics, Reston, VA.
18. ASME Committee V&V 10 (2006), *ASME V&V 10-2006, Guide for Verification and Validation in Computational Solid Mechanics*, 29 December 2006.
19. ASME Committee V&V 20 (2009), *ASME V&V 20, Guide for Verification and Validation in Computational Fluid Dynamics and Heat Transfer.*
20. Wang, S. S. Y., Jia, Y., Roache, P. J., Smith, P. E., and Schmalz, R. A. Jr., eds. (2009), *Verification and Validation of 3D Free-Surface Flow Model*, ASCE/EWRI Task Committee.
21. Oberkampf, W. L., and Roy, C. J. (2010), *Verification and Validation in Scientific Computing*, Cambridge University Press.

22. AIAA (2006), Editorial Policy Statement, January 2006, all AIAA Journals. See also AIAA January 1993, all AIAA Journals.
23. Celik, I. B., Ghia, U., Roache, P. J., Freitas, C. J., Coleman,, H. W., and Raad, P. E. (2008), "Procedure for Estimation and Reporting of Uncertainty due to Discretization in CFD Applications", ASME *Journal of Fluids Engineering*, Vol. 130, No. 7, July 2008, p. 07801.
24. Burton, K., Viceconti, M., Olsen, B., Nolte, L.-P. (2005), "Extracting clinically relevant data from finite element simulations," Editorial, *Clinical Biomechanics*, Vol. 20, p. 451-454.
25. Cowen, R. (2000), "Revved-Up Universe, *Science News*, 12 Feb. 2000, p. 106.
26. Kaiser, D. (2007), "The Other Evolution Wars," *American Scientist*, Vol. 95, Nov.-Dec. 2007, pp. 518-525.
27. Disney, M. J. (2007), "Modern Cosmology: Science or Folktale," *American Scientist*, Vol. 95, Sept.-Oct. 2007, p. 383.
28. Wade, N. (2006), *Before the Dawn: Recovering the Long History of our Ancestors*, Penguin Press, NY.
29. Nagel, E. (1979), *Teleology Revisited and Other Essays in the Philosophy and History of Science*, New York, Columbia University Press.
30. Mehta, U. (1996), "Guide to Credible Computational Fluid Dynamics Simulations," AIAA *Journal of Propulsion and Power*, Vol. 12, No. 5, October 1996, pp. 940–948.
31. Rykiel, E. J. Jr. (1994), in "The Meaning of Models," Letters, *Science*, Vol. 264, No.15, April 1994, pp. 330–331.
32. Rykiel, E. J. Jr. (1996), "The Meaning of Validation," *Ecological Modelling*, Vol. 90, No. 3, pp. 229-244.
33. Refsgaard, J. C. and Henriksen, H. J. (2004), "Modeling Guidelines - Terminology and Guiding Principles," *Advances in Water Resources*, Vol. 27, pp. 71-82.
34. Glanzberg, M. (2006), "Truth," *Stanford Encyclopedia of Philosophy*, http://plato.stanford.edu/*entries/truth/*. June 13.

35. Oreskes, N. (1999), *The Rejection of Continental Drift; Theory and Method in American Earth Science*, Oxford University Press, New York, Oxford.
36. Roache, P. J. (2009), "Perspective: Validation - What Does it Mean?", ASME *Journal of Fluids Engineering*, Vol. 131, No. 3, CID 034503. Also, ASME *Journal of Fluids Engineering* March 2009.
37. Chaisson, E. and McMillan, S. (2008), *Astronomy Today*, Sixth Edition, Pearson Addison Wesley, San Francisco.
38. Deakin, M. A. B. (2007), *Hypatia of Alexandria: Mathematician and Martyr*, Prometheus Books.
39. Riedinger, M. (2010), pers. comm. Feb. 2010.
40. Danielson, D. (2009), "The Bones of Copernicus," *American Scientist*, Vol. 97, Jan.-Feb. 2009, pp. 50-57.
41. Box, George E. P. and Draper, Norman R. (1987), *Empirical Model-Building and Response Surfaces,* Wiley, p. 424.

Chapter 3

Validation:
What Does It Mean?

This chapter is a reprint (plus the addendum Section 3.9), with the permission of the American Society of Mechanical Engineers, of Roache, P. J., "Perspective: Validation - What Does it Mean?," *ASME Journal of Fluids Engineering*, Vol. 131, No. 3, CID 034503. Also, *ASME Journal of Fluids Engineering*, March 2009.

Ambiguities, inconsistencies and recommended interpretations of the commonly cited definition of validation for CFD codes/models are examined. It is shown that the definition-deductive approach is prone to misinterpretation, and that bottom-up descriptions rather than top-down legalistic definitions are to be preferred for science-based engineering and journal policies, though legalistic definitions are necessary for contracts.

3.1 Introduction

> **Validation: The process of determining the degree to which a model {and its associated data} is an accurate representation of the real world from the perspective of the intended uses of the model.**

Unfortunately, considerable disagreement exists on what this definition *means*, or should mean.

This definition of validation has been cited extensively in CFD (Computational Fluid Dynamics) and other computational modeling fields and is widely accepted. Despite the apparent clarity of this concise one-sentence definition using common terms, there is disagreement on its interpretation among scientists and engineers, who are habitually careful readers. There are at least three contested issues: whether *degree* implies acceptability criteria (pass/fail); whether *real world* implies experimental data; and whether *intended use* is specific or general (even by those who think it is needed at all). This gives $2^3 = 8$ possible interpretations of the same definition, without even getting into arguments about what is meant by *model*, i.e. computational, conceptual, mathematical, strong, weak. The job of sorting out claims and arguments is further complicated by the fact that principals in the debates have sometimes switched sides on one or more of these three issues (myself included).

Before examining the definition of validation, we need to make a small distinction on what it is we are claiming to validate, i.e. between *code* and *model*. A model is incorporated into a code, and the same model (e.g. some RANS model) can exist in many codes. Strictly speaking, it is the model that is to be validated, whereas the codes need to be verified. But for a model to be validated, it must be embodied in a code before it can be run. It is thus common to speak loosely of "validating a code" when one means "validating the model in the code," and vendors like to claim they are providing a "validated code," and legal and regulatory requirements may specify use of "verified and validated codes". In theory, the same model would only have to be validated in one (verified) code to be accepted as validated in another (verified) code; in practice for RANS codes, this is unrealistic, so "validating a code" is usually meaningful in context.

3.2 History of the Definition

The definition was precisely stated in a 1996 (re-issued in 2003) U.S. DoD Instruction [1; see also 2,3], which referred to an earlier mini-symposium that used almost the same wording. The DoD re-issue in 2003 [1] added the bracketed additional phrase {*and its associated data*} after the word *model*, which would suggest a strong-sense concept of *model*. The definition was adopted (without the bracketed term) in the AIAA Guide for V&V in CFD [4] and in the ASME V&V 10 [5] which was based in many aspects on [4].[S] The definition is widely used beyond these documents and the observations herein should not be construed simply as criticisms of these sources, but rather as cautions that there are inherent problems with interpretation. The documents cited [1-5] are uneven in their stated interpretations on these issues, with V&V 10 [5] being specific and clear on all three issues. Unfortunately, while acknowledging that a range of definitions exist for validation and other V&V terms, [5] does not acknowledge that a range of interpretations exist for the same definition. Also, while citing [1-4] for its definition of validation, it does not acknowledge the fact that it differs from [1-4] in its interpretation notably on the issue of inclusion of pass/fail criteria. It is also a fact that, for each publication, opinions on what the definition means differ even among members of the same committee that wrote the document. ASME V&V 20 [6] notes the definition but also its range of interpretations, adopting a more general descriptive approach. Likewise, neither the ASCE monograph on V&V for free surface flows [7] nor my 1998 book [8] required the deductive top-down approach implied by legalistic definitions, using instead a descriptive bottom-up approach. It is noteworthy that, in spite of all the agonizing over interpretations, none of the specifics of the complete V&V methodology presented in V&V 20 [6] is affected by any of these choices. Furthermore, while consistent use throughout the computational communities is desirable, there is no necessity for this journal or others to accept a DoD or other definition as canonical, especially when it is easily shown that there are inherent problems with the definition and a wide range of interpretations. (Contracts present a different consideration; see below.)

Based on my contacts in the V&V community, including committee participation in the writing of [5-7], professional contacts with the principals of [4], and experience teaching twelve short courses on V&V, I believe that most professionals make the following interpretations of the definition upon first reading.

3.3 Issue #1. Acceptability Criteria (Pass/Fail)

Regarding the issue of whether acceptability criteria (or adequacy, or pass/fail criteria) are included in this definition of validation, initially people generally say "yes" without hesitation. This is due mostly to a correct recognition that pass/fail decisions must be made in any engineering project, and reinforced by the later phrase "from the perspective of the intended uses of the model" which understandably seems to imply such project-specific criteria (see discussion below). However, people quickly see the value of the alternative view. Although pass/fail criteria are certainly project requirements, the requirements do not necessarily need to be included in the term "validation." In fact, in the original DoD documents [1-3] the term "acceptability" was not used in regard to validation, but in regard to "accreditation" (and which has elsewhere been described as "certification"). From [1]:

> **Acceptability Criteria (Accreditation Criteria). A set of standards that a particular model, simulation, or federation [system of interacting models] must meet to be accredited for a specific purpose.**
>
> **Accreditation. The official certification that a model, simulation or federation of models and simulations and its associated data are acceptable for use for a specific purpose.**

However, acceptability for accreditation as stated in [1] involved additional criteria besides validation accuracy, which supposedly was intended to be included in validation [9]. But close reading of the documents themselves [1-3] give no indication of this, and strongly suggest to me that the acceptability criteria reside under accreditation (or certification, or perhaps another project-related term) rather than validation. The AIAA Guide [4] is somewhat vague (and committee members disagree), and there is widespread misunderstanding of [4] on this point (see discussion below under Issue #3). But V&V 10 [5], even though inspired by the AIAA Guide [4], strongly includes pass/fail criteria, even to the point of insisting that the pass/fail criteria (validation requirements) be set firmly before the comparison to experiment, in the description of intended use. No acknowledgement of this departure from [4] is given in [5], to the likely confusion of any user-engineer who happens to read both documents and who has other things on his mind. It would have been less confusing if the sources each had used different wording for the definition, which might alert the user-engineer, rather than use the same "definition" with different interpretations of the terms.

More important than what the documents state is the fact that people quickly see the advantage of not including a pass/fail tolerance while performing validation. Rather, one simply evaluates the agreement between computational outcomes and experimental outcomes (with their respective uncertainties - see below), and presents the difference as the level of validation. This recognizes the fact that the same validation level (e.g., 10% agreement for skin friction coefficient) may be adequate for one application and not for another. This is just the kind of validation exercise performed for many years for RANS turbulence models, for example.

There are two very distinct processes: first, comparison of model predictions with experimental values, leading to an assessment of model accuracy, and second, determination of acceptability or pass/fail of that accuracy level for a particular application. The methodologies employed in each process have virtually nothing in common. In some usage, a model whose outcomes have been compared to experiments is labeled *validated* regardless of the agreement achieved. In this loosest use of the term, *validated* then is not a quality of the code/model *per se*, but just refers to the QA (Quality Assurance) process. Carried to an extreme, this viewpoint gives the designation *validated* even to very poor models. Celik [10] has pointed out that it would be misleading to assign the inevitably value-laden term "validated" for a code that produces unarguably poor results (say wrong qualitative trends, e.g., lift coefficient decreasing with angle of attack) just because it has gone through the validation QA *process*. I agree, and do not recommend this usage. A more moderate usage is to call the model *validated*, regardless of the agreement achieved, but to state explicitly that the model is validated to a specified level and within the validation uncertainties determined from following the procedures in [6] or other. This way, the validation statement provides a quantitative assessment, but stops short of a rigid pass/fail statement, since that requires consideration of the design, cost, risk, etc. This usage is well presented by Oberkampf et al. [11], p. 348. "Stating our view succinctly: validation deals with quantified comparisons between experimental data and computational data; not the adequacy of the comparisons." The other extreme makes validation project-specific by specifying the error tolerance *a priori*, e.g. see [5]. This ties a code/model validation rigidly to a particular engineering project rather than to less specific science-based engineering (or worse, it neglects the fact that agreement may be acceptable for one application and not for another).

Since not all comparisons should result in a code being given the value-laden designation of *validated*, some minimal agreement should be required. As a reviewer has noted, since it is impossible to avoid attaching a value to *validated*, it can be argued that it is preferable to attach a well defined criterion from the start. But on balance, I think this is outweighed by the disadvantages, as discussed (ephemeral pass/fail criteria, applicability of validation results to more than one project, disparate methods for assessing fidelity and adequacy). The general (and necessarily vague) level of acceptable agreement must be determined by common practice in the discipline. The simulation outcomes with their uncertainties are compared to experiments with their uncertainties, and if reasonable agreement as determined by the state-of-the-art standards (including at least correct qualitative trends) is achieved, then the code/model can be termed *validated*. This does not necessarily mean that the model will be adequate for all applications. Such a project-specific pass/fail tolerance should be relegated to accreditation or certification [8]. The value of this pass/fail tolerance tends to vary over time with design decisions, product requirements, and economics, even though the objective results of the validation comparison itself have more permanent value.

Many discourage the use of the term "validated code" no matter how good the agreement with experiment, because it might be misleading or even deliberately misused, e.g. in commercial code marketing. But it does not seem realistic to try to outlaw the past participle, and codes that have gone through validation will inevitably be referred to as "validated codes." Nevertheless, as Tsang [12, cited in 8, p. 26] noted, "almost by definition, one can never have a Validated computer model without further qualifying phrases." The qualifications include knowledge of the experimental validation set points, the specific validation variables or metrics, what is included in *model*, and of course the degree of validation achieved, which requires stated uncertainties of both computations and experiments.

3.4 Issue #2. Necessity for Experimental Data

In the validation definition, most engineers read "real world" to imply *real world data*, i.e. what most people would call experimental data. Surprisingly, not everyone agrees with this interpretation. (In [1-3] the distinction was not specifically addressed; in [4,5] the requirement was clear and unequivocal, although some members of the committees disagreed.) The apparent motivation is to try to gain the approval implicit in "validation" without the onerous requirement for obtaining real experimental data. There are difficult problems, e.g. nuclear stockpile, for which further testing is outlawed. It is not always clear what these proponents would substitute. Some look for agreement between different models. As noted in [8, p. 276], if one code has been previously validated, it can be regarded as a repository of experimental information, a set of second-hand experimental data plus smoothing and interpolation /extrapolation to parameter values other than experimental set-points. But in general, code-to-code comparison is not validation. The recommended view, agreeing with [5-8], is uncompromising: no experimental data means no validation.

Regarding validation by comparison with a previously validated code, a reviewer has noted that, if a second code is being validated at a set point, the original data could be used, not the first code that has been "validated" at that same set point. In principle, this would usually be the preferred approach. However, for some practical situations the use of a previously validated code could be preferable and certainly more convenient and, I believe, acceptable. First, note that previous multiple validation experiments may not agree with each other, even within the experimental uncertainties (if indeed these have been presented), and they may not be at exactly the same set points. Second, suppose that a new model to be validated is not expected to be as accurate as previous models (but perhaps has an advantage of simplicity, or computational speed, or numerical stability, or lack of sensitivity to grid resolution and is therefore cheaper to run). Then it would make sense to compare the outcomes of the new model with those of a previously validated (but perhaps more complex, slower, less robust, or more expensive) model. (A ready example is a turbulence model using new wall functions, which could be validated against previously validated models employing integration to the wall.) It would be impossible to justify if the new model were intended to be more accurate than the old model taken as a benchmark, except as an interim validation exercise used to justify further validation work (perhaps with new and improved validation experiments).

3.5   Issue #3. Intended Use

The requirement for "intended use" sounds good at first, but it fails upon closer thought. Did D. C. Wilcox [13] need to have an "intended use" in mind when he evaluated the k-ω RANS turbulence models for adverse pressure gradient flows? He may very well have had uses in mind, but does a modeler need to have the same use in mind two decades later? If not, must the validation comparison be repeated? Certainly not.

A Defense of Computational Physics

The "intended use" phrase also bears on pass/fail criteria (Issue #1), seeming to indicate that pass/fail criteria are to be included in the definition of validation. There is widespread misunderstanding of the AIAA Guide [4] on this point, as acknowledged by W. Oberkampf [11], a principal architect of [4]. He states that pass/fail criteria are not included: "We argue that this is what the words mean in the definition ..." The fact that the authors must "argue" the interpretation indicates that the document is unclear, which is understandable given the phrase "from the perspective of the intended uses of the model." Oberkampf insists that "intended use" applies not to a pass/fail tolerance but rather to the metrics involved. Although this observation is relevant, it is not complete, because the same metrics might be applicable to different end uses, just as the same pass/fail tolerances might be. Although [1-4] are not emphatic about specificity of intended use, they are suggestive. V&V 10 [5] is admirably clear but unrealistically strong, even to the point of insisting on *a priori* specification of validation criteria, which if taken seriously would effectively eliminate the possibility of validation in any basic research sense, in my opinion. All these documents [1-5] have a strong orientation to management of large engineering projects, which deters from their applicability to basic research, unlike V&V 20 [6].

Clearly, much of the confusion is the result of trying to use the same word for different needs. Project-oriented engineers are more concerned with specific applications, and naturally tend to rank *acceptability* within *validation* (which term is used more often than *accreditation* or *certification*). Research engineers and scientists tend to take a broader view, and often would prefer to use *validation* to encompass only the assessment of accuracy level, rather than to make decisions about whether that level is adequate for unspecified future uses. It is also significant to recognize that these project-specific requirements on accuracy are often ephemeral, so it is difficult to see a rationale for *a priori* rigid specifications of validation requirements [5,11] when the criteria so often can be re-negotiated if the initial evaluation fails narrowly.

3.6 Recommended Interpretation and Alternative Description

My recommendations, consistent with V&V 20 [6], are that choices for the interpretation of the validation definition be made as follows.

Recommendation on Issue #1. Criteria for *acceptability* of accuracy (adequacy, or pass/fail criteria, or accuracy tolerance) are not part of validation, but analysts performing validation exercises should be wary of appearing to bless a code as "validated" when it is clearly unsatisfactory for any reasonable application (e.g. it cannot even predict correct qualitative trends). In an engineering project, the acceptability of the agreement is part of the next project step, variously called accreditation, certification, or other. It is an engineering management decision, not a scientific evaluation.

Recommendation on Issue #2. Experimental data is necessary for Validation. Many have said unequivocally [5-8,11] that experimental data are the *sine qua non* of validation.

No experimental data => No validation

Many other factors remain, of course, including the quality and quantity of the data, the necessity for uncertainty estimates for both modeling and experiments [6], the extent of the domain of validation (the range of parameter space in the set points of the experiments and the

A Defense of Computational Physics

interpolation/extrapolation of experimental and computational outcomes), whether previously validated codes can be used as a secondary database, whether scaled experiments are adequate, etc. But as a minimum, some experimental data are required. This data can include historical observations and already established scientific facts (especially obvious for invalidation), as pointed out by a reviewer, but it is noteworthy that [4,5] disagree, adopting a literal sense of temporal "prediction" which is at odds with scientific practice.

Recommendation on Issue #3. Intended use, at least in its specific sense, is not required for validation. The common validation definition could be salvaged by re-defining *intended use* to include very general intentions, but frankly this appears to be a hollow exercise. The fact is that a useful validation exercise does not necessarily require an intended use, specific or general. For example, the well-known data on turbulent backstep flow of Driver and Seegmiller [14] in the ERCOFTAC database can be used for code/model validation, with neither the experimenters in 1985 nor modelers in (say) 2008 having a specific use in mind. This is precisely the situation for the Lisbon III Workshop on V&V [15].

However, it is also true and very important that (as recommended strongly in [4-6,11]) experiments designed specifically for a validation exercise, and with a specific application in mind, and with collaboration between experimenters and modelers in the design of the experiments, are much more likely to produce data on the relevant metrics with relevant precisions than are experiments designed without applications in mind.

Alternative Description. Alternately, for science-based engineering, we can *describe* validation rather than rigidly define it. First (and virtually universally agreed upon [4-8,11]) is the distinction between verifications and validation. Verifications (first of the code and then of particular calculations or solutions) are simply matters of mathematics, and address questions of correct coding and discretization accuracy of particular solutions, whereas validation involves comparison with reality, i.e. science (or physics, in its most general sense). In general terms, validation involves comparison of modeling outcomes with experimental results. This has been used in the past, but I agree with [4-6,11] that it is too soft. The trouble (as noted in [11]) is that the difference between model result and experiment is too easily taken to be the accuracy when in fact the story is more difficult. It is time to improve standards somewhat on even the minimal requirements for the term *validation*.

The minimal required improvement is contained in one word: *uncertainty*. We can describe validation (legitimate, minimal validation) as the comparison of model results *and their associated uncertainties* with experimental results *and their associated uncertainties*. A specific methodology for this comparison including interpretation of the answers is given in [6] using accepted, well established quantitative techniques for every aspect of the entire process, and using definitions and statistical techniques that are consistent between experimental and modeling methodologies. I believe that such a descriptive approach is all that is needed for science-based engineering and for journal publication standards. In any case, the warning [12] still applies: it is meaningless to talk about "validation" without significant further qualifications.

3.7 Calibration is Not Validation

Whether one takes a definition-deduction approach or a less rigid descriptive approach, it is necessary to be clear that calibration, the adjustment or tuning of free parameters in a model to fit the model output with experimental data, is not validation. (This distinction is emphasized in each of [4-7] but earlier uses [8] often described calibration as just validation for a restricted range of physical parameters.) Calibration is a sometimes necessary component of (strong sense) model development. But this calibration is not to be considered as validation, which occurs only when the previously calibrated model predictions are evaluated against a set of data not used in the tuning [4-8]. There is no value in tuning free parameters to obtain a drag coefficient to match an experimental value, and then claiming code/model validation because the "prediction" agrees with the same experiment. Historically, this has been a common failing of free-surface flow modeling projects [7]. Of course, if all point-values and functionals of interest are well matched using a small set of free parameters with physically realistic values, this will tend to be convincing in itself, but another data set not used in the tuning will be more so.

3.8 Implications for Contractual and Regulatory Requirements

Although bottom-up descriptions of validation may be adequate for research journals, rigid and legalistic definitions will be required for contracts specifications and regulatory requirements. If a contract specifies that a "validated code" must be used in the modeling, then all parties must know what is meant by "validation" as well as verification, accreditation, etc. My preferences for the definition interpretations are given above, but whatever the contracting or regulating body decides, what is clear from the history of this controversy is the following. Although a rigid, legalistic definition may be required, it is not sufficient. As with questions of constitutional law, interpretations will differ. No matter how carefully the words are crafted, one cannot expect all readers to make the same interpretations.

To better ensure that the intent is correctly interpreted, the contract or regulation specifications should amplify the definitions used with specific interpretations. For example, if the above definition is adopted, the specifications should not just say "real world" and expect the analyst or contractor to know that experimental data is required. The bare legalistic definitions should be expanded to describe the definition, as done notably in V&V 10 [5]. The definition - deduction approach alone is not adequate; the human capacity for equivocation assures that no legalistic definition is inviolable.

3.9 Addendum: Expanded Definition of Validation

As noted, although I prefer descriptions to definitions for technical terms, there often occurs a need for a classical "atomistic" or legalistic one-sentence definition. The following is a suggested definition that expands the widely accepted definition cited at the beginning of the Introduction to include two of the recommended interpretations as discussed previously.

A Defense of Computational Physics

Original Definition

> *Validation: The process of determining the degree to which a model {and its associated data} is an accurate representation of the real world from the perspective of the intended uses of the model.*

Expanded Definition

> *Validation: The process of determining the degree to which a model with its associated data is an accurate representation of the real world as determined by experimental data, the metrics of which are chosen from the perspective of the intended uses of the model.*

This expanded definition still allows interpretation about the adequacy of the agreement for a particular project, i.e. whether or not Validation implies a pass-fail decision. My preference and reasons for not including adequacy in the definition are given earlier, but are not made explicit in this expanded definition. It seems that allowance must be made for the fact that a significant portion of the computational physics and engineering community will continue to include adequacy in the concept of Validation, rather than relegating it to the next step in the process, often referred to as Certification or Accreditation.

Patrick J. Roache

Acknowledgements for Chapter 3

The author gratefully acknowledges discussions with H. W. Coleman and S. S. Y. Wang, and improvements suggested by four reviewers. This work has been partially supported by Idaho National Laboratories, Battelle Energy Alliance Contract no. 54525.

References for Chapter 3

1. Department of Defense, 1996, *DoD Modeling and Simulation (M&S) Verification, Validation, and Accreditation (VV&A)*, DoD Instruction 5000. 61, April 29, 1996. Re-issued 13 May 2003.
2. Department of Defense, 1994, *DoD Modeling and Simulation (M&S) Management*, (DoD Directive 5000.59, January 4, 1994.
3. Military Operations Research Society, 1994, *Simulation Validation (SIMVAL) 1994*, Mini-Symposium Report, 28-30 September 1994.
4. AIAA, 1998, *Guide for the Verification and Validation of Computational Fluid Dynamics Simulations*, AIAA G-077-1998, American Institute of Aeronautics and Astronautics, Reston, VA.
5. ASME Committee PTC-60, 2006, *ASME V&V 10. ASME Guide on Verification and Validation in Computational Solid Mechanics*, 29 December 2006.
6. ASME Committee PTC-61, 2008, *ASME V&V 20. ASME Guide on Verification and Validation in Computational Fluid Dynamics and Heat Transfer*, (expected) 2009.
7. ASCE/EWRI Task Committee, 2009, *3D Free Surface Flow Model Verification/ Validation*.
8. Roache, P. J., 1998, *Verification and Validation in Computational Science and Engineering*, Hermosa Publishers, Albuquerque, New Mexico.
9. Pace, D. K., 2008, pers. comm., 23 May 2008.
10. Celik, I., 2006, pers. comm.
11. Oberkampf, W. L., Trucano, T. G., and Hirsch, C., 2004, "Verification, Validation, and Predictive Capability in Computational Engineering and Physics," *Applied Mechanics Reviews*, Vol. 57, No. 5, pp. 345–384.
12. Tsang, C.-F., 1991, "The Modeling Process and Model Validation," *Ground Water*, November-December 1991, pp. 825-831.

13. Wilcox, D. C., 2006, *Turbulence Modeling for CFD*, DCW Industries, La Canada, CA.
14. Driver, D. M. and Seegmiller, H. L., 1985, "Features of a Reattaching Turbulent Shear Layer in Divergent Channel Flow," *AIAA Journal*, Vol. 23, No. 1, pp. 163-171. ERCOFTAC Classic Database C-30.
15. Eça, L. and Hoekstra, M., 2007, Announcement for the *3rd Workshop on CFD Uncertainty Analysis*, Lisbon, 23-24 October 2008, Instituto Superior Técnico, Lisbon. 1 October 2007.

Glossary

AIAA	American Institute of Aeronautics and Astronautics
ANSI	American National Standards Institute
ASME	American Society of Mechanical Engineering
EWRI	Environmental and Water Resources Institute
ASCE	American Society of Civil Engineering
ASEE	American Society for Engineering Education
APS	American Physical Society
AGU	American Geophysical Union
CFD	Computational Fluid Dynamics
IUTAM	International Union of Theoretical and Applied Mechanics
PDE	partial differential equation
SIAM	Society for Industrial and Applied Mathematics
V&V	Verification and Validation

V&V 10 (a) *ASME V&V 10-200,. Guide for Verification and Validation in Computational Solid Mechanics*

V&V 10 (b) ASME Committee *ASME V&V 10 (formerly PTC-60) on Verification and Validation in Computational Solid Mechanics*

V&V 20 (a) *ASME ANSI Standard V&V 20-2009, Guide for Verification and Validation in Computational Fluid Dynamics and Heat Transfer*

V&V 20 (b) ASME Committee V&V 20 on Verification and Validation in Computational Fluid Mechanics and Heat Transfer

Patrick J. Roache

V&V 30   ASME Committee V&V 30 on
         Verification and Validation in
         Computational Nuclear System
         Thermal Fluids Behavior

A Defense of Computational Physics

Author Vita

B.S. (1960), M.S. (1962), Aeronautical Engineering,
Ph.D. (1967), Aerospace Engineering,
University of Notre Dame

For full resume, visit *www.hermosa-pub.com/hermosa*.

Dr. Patrick J. Roache's primary area of expertise is in the numerical solution of partial differential equations, particularly those of fluid dynamics, heat transfer, and electrodynamics, with special interest in Verification and Validation. He is the author of the original (1972) CFD book *Computational Fluid Dynamics* (translated into Japanese, Russian, and Chinese), the monograph *Elliptic Marching Methods for Domain Decomposition*, the widely referenced *Verification and Validation in Computational Science and Engineering*, the successor to the original CFD book *Fundamentals of Computational Fluid Dynamics*, and the successor to the original V&V book *Fundamentals of Verification and Validation* (2009). He co-edited an ASME symposium proceedings on *Quantification of Uncertainty in CFD*, wrote a chapter on that subject for *Annual Reviews of Fluid Mechanics*, and co-authored a Chapter in the *Handbook of Numerical Heat Transfer*, 2nd Edition. He has authored over 100 archive journal and conference publications, 2 widely referenced reviews, and 20 company reports.

His research experience includes solution-adaptive grid generation, multigrid methods, pseudo-spectral methods, modified method of characteristics, and direct methods for elliptic equations. He has developed high-order time-dependent methods for strongly transient flows and efficient semidirect methods for steady flows. With Prof. S. Steinberg, he pioneered the use of computer Artificial Intelligence

Patrick J. Roache

(Symbolic Manipulation) in CFD and variational grid generation. He developed the Tetra-ELF codes for calculation of nonlinear electric fields in lasers. Other research interests include unsteady aerodynamics, ocean modeling, domain decomposition, groundwater flow and transport, multiphase flow in porous media, passive solar energy system modeling, free convection flows, double diffusion flows, boundary layer transition, base pressure prediction, flow visualization, analytical flight mechanics, and engineering design optimization. He and his staff at Ecodynamics were instrumental in the Performance Assessments for the DOE WIPP (Waste Isolation Pilot Project).

Dr. Roache has extensive teaching experience, including aerodynamics, gas dynamics, boundary layer theory, flight mechanics, heat transfer, fluid dynamics, mathematics (analysis), numerical methods, and CFD. He has taught eleven short courses (six for AIAA) on Verification and Validation. He has served as Adjunct Faculty and Visiting Professor in engineering and mathematics at six universities. He has served as a consultant to the Los Alamos National Laboratories and Sandia National Laboratories, U.S. Army BRL, Kozo Keikaku of Japan, Flow Industries, U.S. Army Corps of Engineers, Air Force Weapons Laboratory, SAIC, LATA, RDA, BDM, and Idaho National Laboratories.

Dr. Roache served as Associate Editor for Numerical Methods for the *ASME Journal of Fluids Engineering* from 1985 to 1988 and co-authored that journal's innovative Policy Statement on the Control of Numerical Accuracy. He also chaired the AIAA Fluid Dynamics subcommittee on Publication Standards for Computational Fluid Dynamics which produced the AIAA Policy Statements on Numerical Accuracy. He has served on the Advisory Editorial Board of six international journals

A Defense of Computational Physics

He has served on Review Boards and Committees for Computational Needs at the U.S. Army Ballistic Research Laboratories, for the New Mexico Energy Institute, for the 1980-1981 AFOSR-HTTM-Stanford Conference on Complex Turbulent Flows, and for the Maui High Performance Computing Center. He chaired a panel to review atmospheric modeling proposals for the U.S. Environmental Protection Agency. He served as Task Leader for CFD for the DOE WIPP (Waste Isolation Pilot Plant) Performance Assessment.

He has received career awards from the University of Cincinnati and the University of Notre Dame. He received the 1994 Knapp Award of the Fluids Engineering Division of ASME for his paper presenting the concept of the Grid Convergence Index, now an established standard method for Solution Verification.

Committee work and publications on Verification and Validation include *ASCE Free Surface Flow Model Verifications*, and ASME Committees on *V&V in Computational Solid Mechanics (V&V 10)*, *V&V in Fluid Dynamics and Heat Transfer (V&V 20)*, and *V&V in Computational Nuclear System Thermal Fluids Behavior (V& 30)*. Both V&V 10 and V&V 20 have resulted in ASME publications accepted as ANSI Standards.

Professional memberships include ASME (Fellow), AIAA (Associate Fellow), and (previously) EWRI, Sigma Xi, Philosophy of Science Association, ASCE, ASEE, APS, AGU, and SIAM.

Patrick J. Roache

A Defense of Computational Physics

End Notes
Chapter 1

[A] Richard von Mises was quoting, in his 1928 classic *Probability, Statistics, and Truth*, Lichtenberg the "natural philosopher" or in modern terms, a scientist, from 1853.

[B] The present author's preoccupation with issues of quality and standards in computational physics (mostly CFD) is evidenced by the following activities.

1985-2012: On the specific subject of V&V, wrote 3 books and 3 other book chapters, co-edited one Proceedings, co-authored 3 committee books (including 2 ANSI Standards), co-authored 2 journal policy statements, taught 11 short courses, wrote 37 papers, and made 41 seminar and meeting presentations.

1972: Published the first book titled *Computational Fluid Dynamics*. Subsequently translated into Japanese, Russian, and Chinese.

1975: Letters column of the AIAA magazine *Aeronautics and Astronautics*, challenging the controversial article of Chapman, Mark and Pirtle entitled "Computers vs. Wind Tunnels in Aerodynamic Flow Simulation" for being too optimistic about numerical approximations.

1979: With G. de Vahl Davis and I. P. Jones, devised a comparison problem for a community assessment of numerical accuracy. "Natural Convection in an Enclosed Cavity: A Comparison Problem," published in *Computers and Fluids*, Vol. 7, pp. 315-316, 1979.

1982: Keynote Speaker for the IAHR Workshop on Numerical Accuracy held in Rome. Devised the benchmark solution for the expanding channel problem.

1985-1988: First Associate Editor for Numerical Methods, ASME *Journal of Fluids Engineering*.

1985: Published (with S. Steinberg), the paper "Symbolic Manipulation and Computational Fluid Dynamics," in *Journal of Computational Physics*, describing in detail what is now known as the Method of Manufactured Solutions for multidimensional code Verification.

1986: Co-authored (with K. Ghia and F. White) the innovative "Editorial Policy Statement on the Control of Numerical Accuracy" for the ASME *Journal of Fluids Engineering*. This was the forerunner of several policy statements addressing the same subject by other journals.

1986-1996: Dr. Roache and his staff at Ecodynamics were instrumental in the annual Performance Assessments for the DOE WIPP (Waste Isolation Pilot Project).

1992-1993: Chaired the AIAA Fluid Dynamics subcommittee on Publication Standards for Computational Fluid Dynamics, which produced the first AIAA Publication Policy Statements on Numerical Accuracy.

1993: Co-edited (with I. Celik., C. J. Chen, and G. Scheurer) the *Proceedings of the Symposium on Quantification of Uncertainty in Computational Fluid Dynamics*, American Society of Mechanical Engineers, ASME FED-Vol. 158.

1994: Published the GCI method (Grid Convergence Index) in the *ASME Journal of Fluids Engineering*. "Perspective: A

Method for Uniform Reporting of Grid Refinement Studies," Vol. 116, No. 3, Sept.1994, pp. 405-413. This paper received the ASME *Robert T. Knapp Award*. "This award recognizes the best paper dealing with results from analytical or laboratory research that has been presented to the Fluids Engineering Division of the ASME within the last two years." The method has subsequently been endorsed (not to the exclusion of possible alternatives, but as a recognized acceptable approach) by the ASME *Journal of Fluids Engineering*, by the ASME publications V&V 10 and V&V 20 (both ANSI Standards), and by AIAA.

1995-2003: Taught eleven short courses on V&V for AIAA, Los Alamos National Laboratories, NASA-Langley Research Center, NASA-Glenn Research Center, University of Notre Dame, ASCE, Canadian CFD Society.

1997: Wrote a chapter on *Quantification of Uncertainty in Computational Fluid Dynamics* for *Annual Reviews of Fluid Mechanics* (Vol. 29, pp. 123-160.)

1998: Published the first book on *Verification and Validation in Computational Science and Engineering*.

1998: Published the book *Fundamentals of Computational Fluid Dynamics*, the successor to the first (1972) book entitled *Computational Fluid Dynamics*, including two chapters on Verification and Validation.

1998: Provided informal critique of the draft AIAA *Guide for the Verification and Validation of Computational Fluid Dynamics Simulations*.

1998-2006: Member, ASCE/EWRI Task Committee on 3D Free Surface Flow Model Verification/Validation Monograph.

2000-2003: Member, ASME/IUTAM Committee on Verification and Validation.

2000-2006: Member, ASME Committee V&V 10 (formerly PTC-60) on Verification and Validation in Computational Solid Mechanics. 2006-2007, Corresponding Member.

2001-2004: Member, Subcommittee of the ASME CFD Technical Committee, subcommittee for Publication Standards.

2002: Published a widely referenced paper on "Code Verification by the Method of Manufactured Solutions", ASME *Journal of Fluids Engineering*, Vol. 114, No. 1, March 2002, pp. 4-10.

2004-2007: Member, ASME Fluids Engineering Division CFD Standards Committee.

2004-2012: Member, ASME Committee V&V 20 (formerly PTC-61) on Verification and Validation in Computational Fluid Mechanics and Heat Transfer.

2004, 2006, 2008: Participated in the three *Lisbon Workshops on CFD Uncertainty Analysis* as Keynote Speaker, member of the organizing committee, and co-author (with L. Eça and M. Hoekstra) of three papers on the workshop results.

2006: Co-authored (with D. Pelletier) Chapter 13, *Verification and Validation of Computational Heat Transfer*, of *Handbook of Numerical Heat Transfer*, Second Edition,

W. J. Minkowycz, E. M. Sparrow, and J. Y. Murthy, Eds., Wiley, New York.

2006-2012: Review Committee, Idaho National Laboratory, Verification and Validation for the Very High Temperature Reactor.

2008: Published book *Fundamentals of Verification and Validation*.

2010-2012: Member, ASME Committee V&V 30 on Verification and Validation in Computational Nuclear System Thermal Fluids Behavior.

2010-2011: Member Organizing Committee, International Workshop on Verification and Validation in Computational Science, 17-19 October 2011, University of Notre Dame.

2012: Keynote Speaker, ASME Verification and Validation Symposium, Las Vegas, Nevada.

End Notes
Chapter 2

[C] S. O. Hansson, pers. comm. July 2010.

[D] However, I would claim that many of today's evolution-based explanations in biology commentaries constitute pseudo-science. The theory is fine, but the practice involves unfalsifiable and glib explanations of virtually any observable fact. Such phony all-powerful explanatory ability is exactly what Popper astutely perceived in Freudians, Adlerians, Marxists and astrologers. For examples, see almost any issue of *Science News*.

E   Kasser (pers. comm. 26 March 2007) offered the following pointed observation. The fact that astrology *is nonsense* is more important than whether it is *scientific nonsense* or *pseudo-scientific nonsense*.

F   Sven Ove Hansson is professor of philosophy and chair of the Department of Philosophy and History of Technology at the Royal Institute of Technology (KTH) in Stockholm, Sweden.

G   In one article, the hypothesis was that cosmic background X-ray radiation originates in previously unknown galactic nuclei. The authors succeeded in finding such discrete sources and showing that these account for a large portion of the background radiation. Contrary to Popper's prescription, this positive finding is much more conclusive than a negative outcome; if no discrete sources had been found, one might conjecture that this was attributable to insufficient resolution.

H   This one case of 70 was a study of certain Hawaiian coastal gravel deposits with the hypothesis that they had been laid down in single events by giant tsunamis. The study showed significant stratigraphic ordering indicating multiple depositional events, thus falsifying the hypothesis in the style of Popper's prescription.

I   It is possible in some situations to perform experiments with (essentially) continuously varying parameters and data acquisition, e.g. continuously varying the angle of a model in a wind tunnel. At least in the one dimensional parameter space of angle, no interpolation of experimental results would be required. Each computational model simulation would occur at a discrete parameter set-point, but the accuracy of the interpolations of these might conceivably be bounded rigorously. (Such bounding would involve consideration of the smoothness of computational solutions, etc. Note that this is a purely mathematical issue, not physical.) Then one could claim Validation accuracy for a continuum without involving unjustified interpolation. But in most cases, experimental data and therefore Validation assessment comes in discrete packages.

J If the only engineering interest were at the experimental set-points, ostensibly one could use the experimental results rather than perform the computations. However, there are other issues. One could have two experiments at identical or nearly identical set-points with large discrepancy, and the computation could be used to decide between them. Or one could have a set of ten experimental set points that includes one outlier; if the computations agreed with the other nine and disagreed with the outlier, one might have more confidence in the computation than in the experiment at the outlier set-point. Most importantly, the experiments may not measure all quantities of interest. The experiments could then be used to Validate the model in the measured quantities, establishing some confidence in the unmeasured quantities of interest. However, whether or not one could then strictly claim use of a "Validated Code" (in those unmeasured quantities) is debatable.

K I am indebted to an anonymous reviewer for some of this wording.

L This example illustrates the fundamental distinction between technical (I claim, normative) modeling use [15-24] of the term Validation and that of Oreskes et al. [6]. They state "Two or more constructions that produce the same results may be said to be empirically equivalent ... If two theories (or model realizations) are empirically equivalent, then there is no way to choose between them other than to invoke extra-evidential considerations like symmetry, simplicity, and elegance, or personal, political, or metaphysical preferences." This statement demonstrates, I believe, that their fundamental concern, like Popper's, is with *Truth* rather than the modest goal of simple accuracy as we use the term herein (others would say "empirical adequacy" [6]). In our view [15-24], there is no *need* to "choose between" alternative formulations, at least not for the process of Validation. However, Oreskes et al. [6] did say the following. "In contrast to the term verification, the term validation does not necessarily denote an establishment of truth (although truth is not precluded)." This statement agrees with our [15-24] intentional meaning of Validation, but not (I think) with their first statement cited.

M   Consider other examples. The non-relativistic compressible Navier-Stokes equations of fluid dynamics are built on Newton's laws of motion, conservation of mass, and continuum flow, all of which are "false" compared to deeper levels of physics. We can further simplify by assuming incompressibility and ignoring viscosity, which are "false" even at an obvious, human scale. From there we can develop the highly simplified laws of vortex dynamics. Not only do these patently "false" laws describe vortex interactions with useful accuracy, they also describe the interactions in a cause-and-effect manner that is not accessible to the deeper levels of physics nor even to the Navier-Stokes equations. I assert that these simplified equations contain *Truth*: this really is how our world works.

This assertion contrasts with Popper's view of *Truth* and of how science proceeds. Consider Popper [1], p. 277, footnote. "... I mean by a crucial experiment one that is designed to refute a theory (if possible) and more especially one which is designed to bring about a decision between two competing theories by refuting (at least) one of them - without, of course, proving the other." Clearly he is concerned with only "truth" in a hierarchical sense, without ever actually *reaching* "truth". Although he acknowledges that new theories will contain the old ones at least to an approximation (a concept he accepts, thankfully), he apparently does not value the old ones once they are "refuted". Such replacement is in fact often a goal of physics, but not, I maintain, the *only* goal of physics. Not only engineers, but most physicists, chemists, etc. concern themselves with other than the ultimate theories. Popper defined progress in science ([1], p. 277) only in terms of new theories which, though unprovable, still supersede old theories. He described this process of "quasi-inductive evolution of science" with an analogy to particles going out of suspension in a fluid, the testable theories settling down in layers of universality, "every new layer corresponding to a theory more universal than those beneath it."

This appealing analogy certainly works for some sequences of theory, e.g. the inclusion of Newtonian mechanics within relativity, but not for all. Caloric theory of heat is not included in modern thermodynamics. (Nor is caloric theory pseudo-science, as sometimes claimed; it is simply incorrect.) This view of science progress has no room for an advance that would be *less* universal, e.g. a new theoretical simplification which is easier to apply, and even provides more insight into the way the world works, while still giving adequate accuracy. Such an approximation would be an *advance*, not a compromise or capitulation. Imagine beginning with tracking every molecule in a contained gas (or perhaps starting earlier with intra-molecular vibrations), advancing to Boltzmann's laws based on statistics, and finally advancing to the ideal gas law $PV = RT$ (or Boyle's and Charles' laws). These would indeed be advances in any practical sense of utility but would constitute a loss of universality. Aside from the possible taint of "applications," even a purist scientist would have to admire the insight into how the world works, and the economy of expression, in PV = RT vs. molecular dynamics. I assert that this theory contains *Truth*, not merely adequate accuracy and convenience.

Indeed, Why stop at molecules? Why not get into electron orbits, or sub-atomic particles, or quantum theory? Answer: Because all this stuff lives at an inappropriate scale. A key concept, seemingly missing from much of Popper's writing, is that of appropriate scale.

The flat earth is really a pretty good theoretical model. If one were calculating the ballistics trajectory from a B-B gun, one would not need a spherical earth, nor higher harmonics of gravity, nor relativity, nor quantum physics, nor string theory. "Advance in science" does not equate with "higher universality."

N   Precisely, Oreskes wrote [35, p. 314] that "Classical mechanics can be partly rescued from the scrap heap of discarded knowledge by arguing that it remains approximately true at speeds far less than [that of] light..." and (p. 368) "Either space and time are absolute or they are not; these are not epistemologically reconcilable positions." (We would use "accurate" rather than "approximately true," a philosophically troublesome term which would seem to dilute any claim to epistemology.) Whatever the resolution of this philosophical point, this fascinating book should be required reading for any student of science.

O   See also Section 9.2.3 and Appendix C of [15] for expanded critiques of [6] and [7]. Besides *confirmation*, Oreskes et al. [6] also used Popper's term *corroboration* and *correspondence*. They also stated the following. "What typically passes for validation and verification is at best confirmation, with all the limitations that this term suggests." But it is our [15,32] opinion that, in common use, "confirmation" does not connote such limitations, although "partial confirmation" would. Oreskes et al. [6] also state the following. "Confirmation is a matter of degree. It is always inherently partial." Again, we see no such automatic connotation of the word "confirmation," and we insist that our explicit explanation of the use of the term Validation is similar to what they claim for *confirmation*. See discussion in Ch. 3 herein, especially Section 3.3, including this statement: "almost by definition, one can never have a Validated computer model without further qualifying phrases."

[P] Often termed "quantities of interest" or sometimes "metrics." [15-23]. I use "metrics" which is easily justified from common use, but the term is sometimes reserved for Validation quantities that functionally include discrepancies and uncertainties [21]. See also [19] and Ch. 11 of [16].

[Q] There is a subtle additional point involved [Riedinger, 39]. Astronomers already knew that Venus was "tied to" the sun, i.e. always close to the sun. If it were not, then the Ptolemaic model would also exhibit all phases.

[R] For a fascinating and readable review of the often-claimed religious preference for geocentricism, see Danielson [40], "The Bones of Copernicus." The theme of displacing mankind's egocentric self-image as center of the universe was not initiated by Copernicus or Galileo or their contemporary natural philosophers, but by the satirist Cyrano de Bergerac (1619-1655), writing more than a century after Copernicus and a generation after Galileo. (This author was the real, historical Cyrano: dramatist, novelist, brave free thinker, and yes, big-nosed flamboyant duelist, and perhaps more pertinent, the pioneering and ardent atheist.) Later, others picked up on the theme, notably the science popularizer Bernard de Bouvier de Fontenelle and the incomparable German poet, philosopher and scientist Johann Wolfgang von Goethe.

For example, Goethe claimed that Copernicus (as proven by Galileo) obliged Earth "to relinquish the colossal privilege of being the center of the universe." Nice imagery, but anachronistic. Goethe wrote this in 1810; Galileo had published in 1610. Galileo would not have related to Goethe's perspective. The early post-medieval European world of Galileo held no such notion. They did not think of Earth as the "center of the universe" but as the pits, the *basement* of the universe. Low and lowly, like a sump pump. The good stuff was up above, like the imagined "fifth element" or quintessence; the dross and stagnant was here on Earth. Giovanni Pico in 1486 had referred to Earth as occupying "the excrementary and filthy parts of the lower world." But Galileo felt that the Copernican view *elevated* mankind's self-image. He explicitly wrote that Earth is no longer "excluded from the dance of the stars. For ... the earth does have motion ... it surpasses the moon in brightness, and ... it is not the sump where the universe's filth and ephemera collect." Also in the early 17th century, Johannes Kepler likewise saw Copernican heliocentricism as a cosmic *promotion* for Earth, not at all the demotion retrojected by Cyrano de Bergerac and later interpreters of the Copernican revolution. The theme continues today as a pseudo-scientific/historic myth. I predict the myth will prove to be too rewarding for moderns to relinquish.

End Note: Chapter 3

[S] Although [5] cited the 2003 version of [1] it did not include the bracketed term added to the original 1996 issue of [1].

www.ingramcontent.com/pod-product-compliance
Lightning Source LLC
Chambersburg PA
CBHW030905180526
45163CB00004B/1706